INFORMAL EMPLOYMENT
IN THE ADVANCED ECONOMIES

Informal employment, whether the result of tax evasion or working whilst claiming benefit, is an issue rapidly rising to the top of government policy agendas.

Informal Employment in the Advanced Economies challenges many of the popular myths surrounding informal economic activities, offering a radical reconceptualisation of their extent, growth, location and nature as well as evaluating the contrasting policy options open to governments.

This book challenges three popular myths about informal employment. First, it tackles the popular belief that informal employment is growing throughout the advanced economies. Second it challenges the myth that this work is undertaken mostly by marginalised groups such as the unemployed, poor, ethnic minorities and immigrants and in deprived neighbourhoods. Third, it evaluates the dominant view that we should replace informal with formal employment through enforcement of stringent laws and regulations concerning tax evasion, benefit fraud and the contravention of labour laws.

In challenging these views, *Informal Employment in the Advanced Economies* finds that far from a universal informalisation of the advanced economies, different places are pursuing heterogeneous development paths: some areas continue to formalise work and welfare; others are witnessing rapid informalisation of a highly exploitative nature. Examining policy options and their consequences, the authors show that conventional regulatory and deregulatory approaches merely exacerbate inequalities and conclude with a radical alternative solution, grounded in a 'new economics' vision of the future of work and welfare provision.

Colin C. Williams is a Senior Lecturer in Economic Geography at the University of Leicester and **Jan Windebank** is a Senior Lecturer in French Studies at the University of Sheffield.

INFORMAL EMPLOYMENT IN THE ADVANCED ECONOMIES

Implications for work and welfare

Colin C. Williams and Jan Windebank

London and New York

First published 1998
by Routledge
11 New Fetter Lane, London EC4P 4EE

Simultaneously published in the USA and Canada
by Routledge
29 West 35th Street, New York, NY 10001

Typeset in Galliard by
BC Typesetting, Bristol
Printed and bound in Great Britain by
Creative Print and Design (Wales), Ebbw Vale

British Library Cataloguing in Publication Data
A catalogue record for this book is available from the British Library

Library of Congress Cataloging in Publication Data
Williams, Colin C.
Informal employment in the advanced economies: implications for work and welfare/
Colin C. Williams and Jan Windebank.
Includes bibliographical references and index.
ISBN 0–415–16959–3 (cloth). – ISBN 0–415–16960–7 (pbk.)
1. Welfare recipients–Employment.   2. Informal sector (Economics)
3. Public welfare.   I. Windebank, J. (Janice)   II. Title.
HV69.W56   1998
362.5'8–dc21   97-50399

ISBN 0–415–16959–3 (hbk)
ISBN 0–415–16960–7 (pbk)

FOR TOBY

CONTENTS

List of tables and figure   ix

Acknowledgements   x

1 Introduction   1

PART I
Examining informal employment: methods and theory   9

2 Methods of researching informal employment   11

3 Theorising informal employment   26

PART II
Socio-spatial divisions in informal employment   47

4 Employment status and informal employment   49

5 Gender and informal employment   66

6 Ethnicity, immigration and informal employment   82

7 Spatial divisions in informal employment   98

8 Informal employment in developing nations   112

PART III
What is to be done about informal employment?
Evaluating the policy options   129

9 Regulating informal employment   131

10 Deregulating formal employment   147

11 Informal employment and the new economics   158

12 Conclusions: re-placing informal employment in the
    advanced economies   173

    Notes   182
    References   188
    Index   213

LIST OF TABLES AND FIGURE

Tables

1.1   Alternative adjectives and nouns used for informal employment   2
3.1   Participation in informal employment in Quebec City, 1986   35
4.1   Wage rates for informal employment in Quebec region:
by employment status, 1993 and 1985   56
4.2   Percentage of public saying that benefit and tax fraud is never
justified   60
4.3   Distribution of informal employment in West Belfast: by
household composition and employment status   62
5.1   Informal wage rates in three regions of Canada: by gender, 1993   67
5.2   Average hourly wage rates in second jobs, Germany   72
5.3   Structure of time by gender and employment status: by country
(in hours per week)   74
5.4   Change in work time in married couples in eight countries,
1961–90   76
7.1   Estimates of the magnitude of informal employment obtained
through indirect methods, as a percentage of GDP   100
8.1   Share of informal sector in non-agricultural employment,
Latin America, 1990 and 1994   115
8.2   Monthly income distributions for formal and informal sectors
in Lima, 1983   118
9.1   The polarisation of employment between households, OECD
nations, 1983–94   137

FIGURE

7.1   Typology of localities relating to the magnitude and character
of their informal employment   105

ACKNOWLEDGEMENTS

This book is the product of over two decades of combined experience of researching and writing on informal economic activity. During this time, there have been many people who have helped develop and influence our thinking on this subject. On the one hand, incisive and creative comments have been made by a host of anonymous referees who have read our previous writings on this subject. Here, therefore, we wish to thank these unknown people for undertaking in a professional manner this time-consuming academic task that so often goes unrewarded and unrecognised. On the other hand, there are the participants, too numerous to mention, who have actively contributed ideas at seminars and conferences where we have given presentations. Their comments have helped shape our thinking on this subject in ways that often become forgotten as their comments become further developed into 'our' ideas.

During the preparation of the book itself, meanwhile, it has been wonderful to see that despite the increased pressures on academics throughout the world, numerous individuals have been willing to give their time to help us in this venture and that despite pressures for retrenchment, co-operative interdisciplinary and cross-national collaboration is still possible. A wide range of people have provided help and advice in the process of writing this book. Without their aid, we are certain that the learning and writing process involved in this endeavour would have been much slower. In this regard, we should like to thank particularly a number of colleagues who freely gave their time both to read and make incisive comments on draft versions of the chapters, namely Franck Duvell, Bill Jordan, Guy Lacroix and Ed Mayo. We are also grateful to the anonymous readers for Routledge who offered some excellent ideas on the structure of the book that greatly facilitated it becoming a more coherent whole.

In addition, many freely gave their time to provide information and advice, showing that the collective enterprise of the pursuit of knowledge is not yet dead. These people include Barbara Brandt, Hartley Dean, Bernard Fortin, Pierre Frechette, Jonathan Gershuny, Leif Jensen, B. Kazemeier, Annie Komter, Thomas Lemieux, Madeleine Leonard, P. Lysestol, Richard Macfarlane, Anne Oberhauser, Cathy Rakowoski, Piet Renooy, Saskia Sassen and Mark Warren. Of course, the normal disclaimers apply: any omissions or faults are ours alone.

Finally, both of us owe a good deal of gratitude to our son, Toby. We have written this book during the first year of his life. Without his 'wakefulness', we would surely not have had so many hours a day to discuss the contents of this book and neither would we have had the incentive to rise so early so that one or other of us could sit at the computer whilst the other tended to his needs. For that and much much more, we are sincerely grateful to him.

1

INTRODUCTION

One of the main arguments emerging in the social sciences in recent years has been that social and economic restructuring, together with government policies, have led to the creation of a large and growing sphere of paid economic activity beyond the realm of formal employment. Variously referred to as the black economy, the underground sector, hidden work or the shadow economy to cite but a few of its titles, this economic activity involves the paid production and sale of goods and services that are unregistered by, or hidden from, the state for tax, social security and/or labour law purposes but which are legal in all other respects. As such, it includes within its scope not only social security fraud by those 'working on the side', 'doing the double' or 'fiddling', but also the evasion of tax or flouting of labour legislation by 'fraudsters', 'wide-boys', 'spivs' or 'cowboys'. This subject is one where everybody has a story to tell either about their own or others' activities. It is also a subject about which many myths have emerged in popular discourse concerning the size and character of the phenomenon as well as who is doing it and where. In short, these myths suggest that such activity is growing in the advanced economies (the 'informalisation thesis') and that it is undertaken mostly by marginalised groups such as the unemployed, poor, ethnic minorities and immigrants and in deprived neighbourhoods (the 'marginality thesis').

If such views were limited to the occasional scaremongering media story, such hyperbole would be of little concern. The problem, however, is that these perceptions of informal employment have played an important and influential role in shaping not only policy approaches towards such work but also wider economic and social policies. The aim of this book is therefore to evaluate critically these views by analysing not only the magnitude and character of informal employment but also what should be done about this work in the advanced economies. In order to set the scene for what follows, this introductory chapter thus first defines informal employment, second, explains the rationale for this publication and third and finally, outlines the structure of the book.

1

Defining informal employment

What we call informal employment is known by many different names. Table 1.1 provides a list of the alternative adjectives and nouns used to denominate this activity. Throughout Europe, the most common adjective used is 'black'. This is the most popular term employed in the UK and the Netherlands and the second most favoured name in France (following 'subterranean'), Germany (behind 'shadow') and Italy (after 'submerged'). In North America, in contrast, 'black' is not utilised. Instead, 'underground' is the standard adjective employed and 'hidden' the next most popular name (see Thomas 1992).

Following the North American tendency, the adjective 'black' is rejected for the purposes of this book. Although this may be still widely used in Europe, its popularity is quickly waning. This is because to use 'black' as a metaphor for such activity and to contrast it with the legal or formal 'white' economy is fast being recognised as politically incorrect in the context of equal opportunities. Here, therefore, and despite having used 'black' in the past to define such work (Williams and Windebank 1995a), we must search for another adjective. Many of the alternatives in Table 1.1, as we shall see later in this book, are inadequate. Although much informal employment is 'irregular' or 'precarious' for

Table 1.1 Alternative adjectives and nouns used for informal employment

Adjectives	Nouns
Black	Economy
Cash-in-hand	Sector
Clandestine	Activity
Ghetto	Work
Hidden	
Invisible	
Irregular	
Non-official	
Off-the-books	
Other	
Parallel	
Precarious	
Second	
Shadow	
Subterranean	
Twilight	
Underground	
Unobserved	
Unofficial	
Unorganised	
Unrecorded	
Unregulated	

Source: Authors' survey.

example, not all such activity is of this type. Some engage in regular or stable activity but it is their individual status which renders their particular employment 'informal'. Neither are 'unorganised' or 'unregulated' accurate descriptions for reasons stated below and which will be returned to throughout this book. Adjectives such as 'shadow', 'subterranean', 'invisible', 'hidden', 'unobserved', 'underground' and 'twilight', meanwhile, all suggest that such activity somehow exists in the hidden interstices of contemporary society. We reject this metaphor because although mostly hidden from the state for certain limited administrative purposes, such activity is often very visible in the communities in which it takes place (see, for example, Harding and Jenkins 1989). Indeed, in areas where informal employment is condoned, it remains visible even to the state authorities.

One prominent adjective, which remains, therefore, once these others have been rejected, is 'informal'. Indeed, this is the most popular adjective used to describe all forms of paid and unpaid work existing outside employment. It is also perhaps one of the most accurate adjectives which could be used to portray how the social relations in this realm of economic life differ from the more 'formal' social relations in which official employment is embedded. A problem with using this adjective, however, is that when combined with nouns such as 'economy', 'sector', 'work' or 'activity', it is normally seen to include both paid and unpaid work for the majority of commentators. For that reason, we use the adjective 'informal' but attach it to the noun 'employment' since employment denotes a relationship where one's labour is recompensed by a wage or fee, in other words, paid.

By attaching the adjective 'informal' to the noun 'employment', we also overcome many of the other problems which result from combining this adjective with the alternative nouns listed in Table 1.1. To call such activity an 'economy', for instance, not only leads many readers to assume that one is discussing both paid and unpaid work but conjures up the image of 'dual' or 'separate' economies when it has become increasingly apparent that activities are all part of one economy (Cappechi 1989, Harding and Jenkins 1989, Thomas 1992). As Gershuny (1985: 129) asserts, 'the informal economy . . . is of course not a separate economy at all but an integral part of the system by which work, paid and unpaid, satisfies human needs'. Indeed, to speak of work outside of employment as an 'economy' is to imply that such work enjoys a degree of autonomy from the formal market. As this book argues throughout, however, there exists an intimate interdependent relationship between informal employment and its formal counterpart. For this reason, any reference to such employment as an 'economy' has been deliberately avoided throughout this book.

Using the term informal employment also avoids the other common tendency of referring to such paid activity as a 'sector' (e.g. Felt and Sinclair 1992, Thomas 1992). The problem here is that sectors in common parlance and in the Standard Industrial Classification (SIC) index, are defined by the good they produce or the nature of the service offered. Informal employment, nevertheless, cannot and is not defined in such a way. It is not constituted by a particular set of tasks or activities but cross-cuts all sectors. Informality, put another way, is not an inherent property of specific activities (e.g. homeworking, window-cleaning). Instead, whether an activity is informal or not is a social construction (Portes 1994) and all goods and services can be produced and distributed either formally or informally.

Moreover, the use of both 'economy' and 'sector' is here rejected because it is an exercise in residualising all work that is not formal employment. It leaves formal employment intact and defines everything which is not formal employment as the 'other', thus clustering together the many remaining heterogeneous activities ranging from unpaid housework through repairing a neighbour's tap for cash to an employer running large-scale unregistered factories. This reflects the prevailing ideology of advanced economies that employment is the principal form of work whilst all non-employment is put into a residual 'sector' or 'economy'. The result is that not only the difference between paid and unpaid informal economic activity, but also the heterogeneity within informal employment, is hidden from view.[1]

Unfortunately, however, and whatever term is used, it is impossible to escape from defining informal employment in a negative and residual manner. This is because it catches all paid work that is not employment. As such, it cannot be defined in a more positive way. Nevertheless, this should not be seen as a problem, which can be overcome with sufficient intellectual ingenuity. Rather, it must be viewed as a central pillar in understanding the current perception of formal and informal employment in the advanced economies. Informal employment is what formal employment is not and must therefore remain defined in such a manner.

Here, therefore, we use the term informal employment to refer to the paid production and sale of goods and services that are unregistered by, or hidden from, the state for tax, social security and/or labour law purposes, but which are legal in all other respects. As such, informal employment is composed of three types of activity: evasion of both direct (i.e. income tax) and indirect (e.g. VAT, excise duties) taxes; social security fraud where the officially unemployed are working whilst claiming benefit; and avoidance of labour legislation, such as employers' insurance contributions, minimum wage agreements or certain safety and other standards in the workplace, such as through hiring labour off-the-books or sub-contracting work to small firms and the self-employed asked to work for below-minimum wages. This accords with the standard definitions of informal employment in the social sciences literature. On the one hand, it excludes unpaid informal work from its scope, including the production of goods and services for a family's own consumption or as an unpaid favour for friends, neighbours or one's community. On the other hand, it explicitly denotes that the only criminality about informal employment is the fact that the production and sale of the goods and services are not registered for tax, social security or labour law purposes (Feige 1990, Portes 1994, Thomas 1992). Criminal activities where the goods and services themselves are illegal are not included in this definition of informal employment.[2]

Although this definition of the nature and scope of informal employment conforms to most other definitions in the literature (e.g. Castells and Portes 1989, Pahl 1984, Sassen 1989, Thomas 1992), it is important to make explicit that it differs in one important manner. It does not define informal employment as 'the *unregulated* production of otherwise licit goods and services' (Castells and Portes 1989: 15). Following Warren (1994: 93), we replace the term 'unregulated' with the term 'unregistered', 'because the state can regulate, in the sense of shape or influence, the nature of the informal economy, often in unintended ways'. Indeed, the notion that informal employment is unregulated is one of the principal myths that this book sets out to challenge. In the advanced economies, not only does the state regulate the nature and scope of informal employment by changing its rules and regulations towards formal employment but it also often plays an active role in legitimising informal employment through such practices as deliberate lax enforcement or even positive support for this activity. Moreover, informal employment is regulated by other factors besides the state. As will be shown throughout this book, there are a host of economic, social, institutional and environmental regulators, which shape the nature and extent of such activity in any population or place. The problem with defining informal employment as 'unregulated', therefore, is that one is not only erroneously suggesting that such activity constitutes a free market operating independently of social, economic, institutional and environmental influences, but one is also pre-judging the variables which will be relied upon to explain the existence and growth/decline of informal employment in a way which is unconducive to understanding.

Rationale for the study of informal employment

Why study informal employment? For many years informal employment, although its existence was not denied, was assumed to be in decline and/or of little relevance for understanding and interpreting advanced economies and their prospects. Instead, an assumption prevailed that informal employment was a mere leftover from a previous era of production. Work and welfare would become increasingly formalised and result in an end-state of full employment and comprehensive and universal formal welfare provision. However, the history of the past twenty years in the advanced economies and beyond has taught us that such an assumption can no longer be accepted. Rather, a far-reaching restructuring of work and welfare has brought starkly into question the inevitability of continuing formalisation and thus that full employment coupled with comprehensive welfare provision is the natural end-state of our development pathway.

5

Moreover, there is no reason to believe that informal employment is a small-scale marginal activity which is in decline and will disappear as economies become more advanced. Indeed, given the fiscal and welfare problems currently confronting the advanced economies, many governments now believe that they can no longer afford to ignore informal employment. This has led to the emergence of a wave of studies on such work.[3] Faced with empty public purses, tax evasion and social security fraud are issues rapidly rising to the top of government agendas as they attempt to both raise revenue and possibly reduce welfare payments. In this regard, a number of questions have arisen over this form of employment.

First, there are questions over its size and growth. Why does informal employment exist? Is it simply a leftover from a previous economic system which will eventually disappear or does it signify the emergence of a new form of advanced capitalism? Is such employment growing or declining in the advanced economies? Does its existence reduce the prospect of formal jobs being created or is there a complementary relationship between formal and informal employment?

Second, there are questions over the character of informal employment. Who engages in informal employment? Is it mostly populations marginalised from formal employment such as the unemployed, poor, women, ethnic minorities and immigrants? Or is it more complex than this? Why do people engage in such employment? Do they participate out of choice, for example to evade taxes, or is their participation involuntary and a last resort in order to earn income in order to survive? What type of activity do they engage in? Is such employment always low-paid, exploitative and monotonous or can informal employment be highly paid, autonomous, meaningful and rewarding? Where does it take place? And does informal employment reduce or reinforce the existing social and spatial inequalities produced by formal employment?

Third, and finally, there are questions concerning policy towards this form of employment. What sort of futures for work and welfare are we seeking and how does informal employment fit into these futures? Should such employment be formalised and if so, how? Should it be eradicated through tougher regulations? Or should informal employment be used as a model for the rest of the economy? Alternatively, should we adopt a 'whole economy' view of work, which seeks 'full engagement' based on the active promotion of not only formal employment but also unpaid informal work? These are the central questions, which this book sets out to answer.

Structure of the book

In order to answer such questions, this book is divided into three parts. Part I evaluates the methods and theories employed in order to understand informal employment in the advanced economies, Part II is an analysis of the socio-spatial divisions of informal employment and Part III is an examination of what is to be done about such work in the advanced economies.

In Part I, we set the scene for our analysis of the extent and nature of informal employment across different populations by evaluating critically how informal employment has been examined. In chapter 2, therefore, we assess the validity of the various techniques used to evaluate the magnitude and character of such work, ranging from the indirect monetary and non-monetary methods to the more direct survey approaches. This reveals that great caution is required when interpreting the results of surveys, due to the inherent problems in many of the techniques used to study this form of work. With this cautionary note in mind, chapter 3 then sets about the task of synthesising the findings of the vast array of studies of informal employment in the advanced economies so as to theorise first, its magnitude and second, its character. This reveals that informal employment is not a phenomenon extinguished by the modernisation of economic and social life in the second half of the twentieth century (the formalisation thesis) which has again risen, albeit in a different form, from the ashes of economic restructuring since the 1980s (the informalisation thesis). Instead, the idea is introduced that different localities, regions and nations within the advanced economies are pursuing heterogeneous development paths where formalisation and/or informalisation are occurring to differing extents and at different rates. Alongside this recognition of the need to *re-place* informal employment, the idea is also introduced that informal employment is not simply a peripheral form of employment undertaken by marginalised populations as a means of survival (the 'marginality thesis) but is a far more complex entity which is hierarchically structured in terms of access, rewards and experience.

Part II, in consequence, examines the socio-spatial divisions of informal employment in greater depth so as to debunk many of the myths which have arisen out of the marginality thesis. When considering the distribution of informal employment, the *a priori* assumption has often been that marginalised social groups such as the unemployed, women, ethnic minorities and immigrants and those living in deprived areas are more heavily engaged in this employment. In order to examine whether this is indeed the case, Part II evaluates the nature and extent of informal employment undertaken according to employment status (chapter 4), gender (chapter 5), ethnicity and immigration (chapter 6) as well as geographically (chapter 7). All of these chapters are structured in the same manner. First, each reviews the extent of informal employment undertaken by the marginalised population under consideration and second, examines the nature of the informal employment they conduct, their motivations for participating in such work and wage rates. Finally, each chapter seeks explanations for the magnitude and character of the informal employment they undertake. In chapter 8, meanwhile, we consider whether the tendencies in informal employment in the advanced economies are also applicable to the developing nations. To do this, we examine the extent of informal employment in developing nations and its nature according to employment status, gender, ethnicity and migration as well as geographical spread.

Having reviewed the extent and nature of informal employment, Part III then examines the policy options concerning this form of work. Over the past twenty years, two of the key spheres of policy-making in the advanced economies have been on the one hand, the development of economic strategies to help rejuvenate regions and localities suffering from the vagaries of global economic restructuring and on the other hand, the construction of welfare strategies to tackle the problem of social marginalisation. Until now, however, and despite the accumulating knowledge on the role of informal employment in economic restructuring and welfare provision, there has been little attempt to feed this knowledge into policy formulation.[4]

In Part III, therefore, we explore what can and should be done about informal employment in advanced economies. So far as is known, this is one of the first attempts to examine these policy issues in any depth. Here, three possible policy approaches towards informal employment are evaluated. These are derived from a review of the stances adopted, mostly implicitly, in the literature on informal employment. First, there is the 'regulatory' perspective, which advocates the eradication of such work as part of its bid to achieve full employment and a comprehensive formal welfare system. Second, there is the 'deregulatory' perspective, which views informal employment as an indictment of state interference in formal employment and demands more deregulation of the formal labour market as a means of blurring the distinction between formal and informal employment. Third and finally, there is the 'new economics' perspective which advocates the introduction of a 'guaranteed basic income' and assisted self-help so as to restructure radically the modes of both economic production and welfare provision and, in so doing, abolish the necessity for people to undertake exploitative informal employment.

In each chapter, the same format is used. First, the particular approach it adopts towards informal employment will be analysed in the context of its view of the future of work and welfare and second, a critical evaluation is undertaken both of this broader vision of work and welfare as well as the practicality and desirability of its approach towards informal employment. This reveals that the discussion of what can and should be done about informal employment is inextricably connected to the wider debates on work and welfare in the advanced economies. Indeed, and as the sub-title of the book intimates, a principal aim of this book is to show how the study of informal employment, far from being a peripheral subject of inquiry, enables some radical new insights into the validity of the alternative futures for work and welfare currently being propagated in the advanced economies.

Part I

EXAMINING INFORMAL EMPLOYMENT
Methods and theory

2

METHODS OF RESEARCHING INFORMAL EMPLOYMENT

Introduction

In order to lay the foundations for understanding informal employment in the advanced economies, this chapter evaluates the contrasting approaches for measuring the magnitude and character of such activity that have been developed. This reveals the need for great care when analysing and comparing the findings of studies due to the limitations of each technique used. Consequently, it calls for detailed attention to be paid to the methods employed before reading off trends from any results. Unless this is done, all attempts to review the composition and size of informal employment will end in confusion over how to interpret apparently contradictory findings. It is only by unpacking the methods utilised in each study and interpreting their results in terms of the limitations of the methods that a clear understanding of informal employment can be achieved.

Given that informal employment is hidden from, or unregistered by the state for tax, social security and/or labour law purposes, no neat figures exist detailing its size. A major debate concerning whether it should be directly or indirectly measured has ensued. First, there are those who assume that research participants will not be forthcoming about whether or not they engage in such activity. These researchers thus seek evidence of informal employment in macro-economic data collected and/or constructed for other purposes. The belief is that despite informal workers wishing to hide their incomes, their activities will none the less reveal themselves at the macro-economic level and it is these statistical traces of paid informal activity that are sought by indirect approaches. On the whole, these approaches concentrate on estimating the volume or value of informal employment either in monetary terms or in relation to formal employment. Much less emphasis is put on identifying the character of informal employment. Second, and in contrast, are those who assume that despite the illegal nature of the phenomenon, one can rely on the honesty of research participants when they are questioned about their informal employment. These analysts have thus conducted mostly intensive investigations on small samples, such as through locality studies, of the nature and extent of informal employment using direct survey methods.

Here, therefore, we provide an overview and critical evaluation of these contrasting methods of researching informal employment, examining first the relatively indirect methods of the macro-economic approaches and second, the more direct survey methods. At the outset, however, it is important to state that in reality, we are dealing with a spectrum of approaches which sometimes overlap and cross-cut one another.

More indirect approaches

These relatively indirect approaches to evaluating informal employment are of three varieties. First, there are indirect methods which examine non-monetary indicators such as the discrepancies in labour supply figures or the number of very small firms. Second, there are indirect monetary methods, which look for traces of informal employment in economic aggregates such as money supply, and third, there are those methods which examine discrepancies between income and expenditure levels either at the aggregate or household level. Each is considered in turn. The problem with all of these approaches, as will be shown, is to find a reliable and exact relationship between these so-called 'traces' of informal employment and the actual volume of such work.[1]

Indirect non-monetary methods

Two of the most popular indirect approaches which use non-monetary sur- rogate indicators to estimate the extent of informal employment in advanced economies are first, those which look for informal employees in formal labour force statistics and second, those which take very small enterprises as a proxy for the extent of informal employment.

Labour force estimates

Those methods that seek to identify the informal labour force in formal labour force statistics are of two varieties. On the one hand, there are those that have identified various types of employment (e.g. self-employment, second-job- holding) in which workers are most likely to be employed informally and looked for unaccountable increases in the official labour force statistics in the numbers employed in these categories (Alden 1982, Del Boca and Forte 1982). The problem, however, is that the resulting statistics have usually been based on little, if any, evidence of the extent to which these job categories are populated by informal employees and there are a multitude of additional factors besides informal employment which might lead to an increase in these categories. Self-employment, for instance, is not solely the result of a rise in informal employment. Since the early 1980s when the majority of studies using this method were conducted, it has also been seen to be due to such trends as the rise of the enterprise culture, increased sub-contracting in the production process and other forms of flexible production arrangement. Second-job-holding, moreover, is not always directly a product of informal employment except if such job-holding is illegal *per se*. It is also due to the combined effect of broader economic and cultural restructuring processes such as the demise of the 'breadwinner wage' and the proliferation of part-time work. To identify the proportion of the growth in self-employment or multiple-job-holding attributable to such processes and the share attributable to informal employment at a particular moment is a difficult if not impossible task.

On the other hand, there are those who examine discrepancies in two different methods used to compile official employment statistics in order to identify the level of informal employment. In the US, for example, this has entailed comparing the Census Bureau's Current Population Survey (CPS) with the Bureau of Labor Statistics (BLS) survey of firms. The CPS includes a monthly sampling of about 60,000 households in which questions are asked about the work status of their occupants and everyone is classified as employed, unemployed or not in the labour force, whilst the BLS survey examines establishments to determine the number on the payroll. The comparison of the two data sets has been premised on the assumption that those working informally would declare themselves, or be declared, as job-holders in the household survey, but would not show up on the books of business enterprises. The discrepancy in the numbers between the two surveys has been thus taken as the number employed informally, with changes in the difference between the two sets of figures seen as a measure of its growth or decline (see, for example, Denison 1982, Mattera 1985, US Congress Joint Economic Committee 1983). Similar methods have also been adopted in Portugal by the Instituto de Pesquisa Social Daniao de Gois which examines the number of informal employees by looking at the difference between the total number of people registered as salaried workers and the total number of workers registered in the Ministry of Work statistics (see Lobo 1990b).

The problems with such an approach, however, are manifold. In terms of its relevance for measuring the size of the informal labour force, the first problem with this method is that it has erroneously assumed that each individual is either a formal or informal worker and in so doing, missed a vast amount of informal employment conducted by those who have a formal job but also engage in informal employment. Second, by examining only those employed in businesses, it has missed numerous informal workers who undertake jobs for households on an informal self-employed basis. Third, there is no reason to assume that an informal employee will describe him/herself as employed in a household survey whilst the employer will not in a business survey. Finally, the fact that such analyses have resulted in contradictory results, with some studies showing no change in the size of informal employment in the post-war years (e.g. Denison 1982) and others showing growth (US Congress Joint Economic

Committee 1983) gives rise to the need for great caution. Identifying the informal labour force through formal labour force statistics, in sum, has been beset by problems that cannot be easily transcended and this method has waned in popularity since its heyday in the early 1980s.

The 'very small enterprise' approach

For some, very small enterprises (VSEs) have been adopted as an alternative non-monetary proxy of the extent of informal employment (e.g. Portes and Sassen-Koob 1987). The assumption is that in advanced economies, most employment of informal workers occurs in smaller enterprises because of their reduced visibility, greater flexibility and better opportunities to escape state controls. Larger firms, meanwhile, are assumed to be more vulnerable to state regulation and risk-averse to the potential penalties and thus less likely to employ informal employees directly, although they are purported to sub-contract to smaller firms who use such labour.

This approach has been employed as an indicator of informalisation by the Wage and Hour Division of the US Deptartment of Labor, the agency charged with enforcing minimum wages, overtime and other protective codes for US workers. Their interviews revealed widespread violations of the labour codes among garment, electronics and construction sub-contractors as well as in all kinds of personal and household services, especially in large metropolitan areas. Most of the enterprises were very small, composed of less than ten workers, thus reinforcing the notion that informal employment is closely tied to VSEs (Fernandez-Kelly and Garcia 1989, Sassen and Smith 1992). A further study by the US General Accounting Office (1989) identified the restaurant, apparel and meat processing industries – all industries where small firms predominate – as having the greatest incidence of 'sweatshop' practices, such as failing to keep records of wages and work hours, making payments below the minimum wage or without overtime pay, employment of minors, fire hazards and other unsafe work conditions.

As an indicator of informality, however, the VSE approach is subject to two contradictory biases. First, not all small firms engage in informal practices, which could lead to an overestimate; second, fully informal VSEs will escape government record-keeping which could lead to an underestimate (Portes 1994). A further complication is that the extent to which VSEs engage either wholly or partly in informal practices will vary according to the geographical context in which they are operating. As such, estimates of both the size and growth/decline of informal employment can only be very approximate. More widely, it ignores sub-contracting by large companies and more individual informal work conducted by people on a one-to-one basis to meet final demand. As Portes (1994: 440–1) thus concludes,

By themselves . . . such series represent a very imperfect measure of the extent of informal activity. It is impossible to tell from them which firms actually engage in irregular practices and the character of these practices. All that can be said is that small firms, assumed to be the principal locus of informality, are not declining fast and actually appear to increase significantly during periods of economic recession.

Given that such non-monetary proxies are but crude measures of the extent of informalisation, other indirect methods have been based on monetary indicators in the belief that this will enable a much closer estimate of the magnitude of informal employment in the advanced economies.

Indirect monetary methods

Here, three different methods that have searched for indirect monetary evidence of informal employment are evaluated in turn. These are the high-denomination notes, the cash–deposit ratio and the money transactions approaches.

The high-denomination notes approach

Based on the belief that the circulation of high-denomination bank notes is a key indicator of the volume of informal employment (Freud 1979, Henry 1976), this methodology, used principally during the 1970s and early 1980s, but now having fallen out of favour, is embedded in a caricature of informal workers carrying around a fat roll of bank notes. It assumes not only that those in informal employment use cash exclusively in their transactions but that they handle large quantities of cash and exchange high-denomination bank notes. Henry (1976), for example, argues that the increasing demand for $50 and $100 notes in the US between 1960 and 1970, a period which saw a rapid increase in non-cash methods of payment such as personal cheque accounts and credit cards, could only be explained in terms of an expansion in profit-orientated crime and tax evasion. He asserts that these activities require extra cash so as to avoid leaving traceable records. Taking into account factors such as price levels, personal consumption expenditures and federal income tax revenues, he estimated that the extra demand for high-denomination notes (US$50 and over) resulting from tax evasion to be as high as US$30 billion in 1973.

However, whether this method can be used either to estimate the magnitude of informal employment or as evidence of a growing informalisation of the advanced economies is doubtful. First, this method has no means of separating the use of high-denomination notes in crime from their use in informal employment, meaning that there is no means of knowing what proportion of the circulation of high-denomination bank notes should be taken as an indicator of high levels of crime and what share is indicative of informal employment. Second, there is not even a strong rationale for assuming that tax evasion is paid for in high-denomination notes. Indeed, many informal trans-actions, as will be shown in later chapters, are for relatively small amounts of money (e.g. Cornuel and Duriez 1985, Evason and Woods 1995, Tanzi 1982) and do not necessarily even involve the exclusive use of cash in exchange (see below). Measuring such activity in terms of the use of high-denomination bank notes, therefore, may well be a poor indicator of the level of informal employ-ment.

Even if these problems could be overcome, there are then a range of prob-lems inherent in utilising this method as evidence of the growing informalisa-tion of the advanced economies. Although the number of high-denomination notes has increased in advanced economies, the overwhelming finding is that this cannot be taken as evidence of informalisation. First, there is inflation to be taken into account. Porter and Bayer (1989) in the US present strong econo-metric evidence for a relationship between per capita holdings of $100 bills and the price level. In the UK, meanwhile, the increase in high-denomination notes has been less than the rate of inflation. Between 1972 and 1982, the retail price index rose by 290 per cent whilst the average value of the denomination of bank notes only rose by 120 per cent, meaning that the average denomina-tion has declined by 40 per cent over this period when inflation is taken into account (Trundle 1982). In Canada, moreover, Mirus and Smith (1989) note little change in real terms. Second, since this approach was propagated, there have been profound transformations in both attitudes and behaviour towards cash payments. On the one hand, there have been major alterations in modes of payment (e.g. credit and debit cards, store cards) resulting in a decline in cash usage. On the other hand, there has been a restructuring of formal finan-cial services in the advanced economies, reflected in the 'flight of financial institutions' from poorer populations (Leyshon and Thrift 1994) and resulting in increased cash usage amongst the financially excluded. These represent counter-tendencies which make it difficult to discern whether changes in cash usage are due to the restructuring of formal financial services, shifts in attitudes and behaviour, or the growth/decline of informal employment. The high-denomination notes method, in sum, represents a very unreliable indicator of either the extent or changing magnitude of informal employment.

The cash–deposit ratio approach

Instead of examining the volume of high-denomination bank notes, another indirect monetary method has been to measure the ratio of currency in circu-lation to demand deposits. Again based on the assumption that in order to conceal income, informal transactions will occur in cash, this approach consists of arriving at an estimate of the currency in circulation required by the opera-tion of legal activities and subtracting this figure from the actual monetary mass. The difference, multiplied by the velocity of money, provides an estimate of the magnitude of the 'underground economy'. The ratio of that figure to the observed GNP then gives the proportion of the national economy represented by underground activities.

Indeed, it was this approach, developed by Gutmann (1977, 1978), that brought informal employment to the attention of the public in the US. He estimated that this work was worth some US$176 billion in 1976. Making the heroic assumption that there was no informal employment in the US prior to World War II since levels of taxation were so low as to make tax evasion strategies unnecessary, he takes the ratio for this period as the baseline norm and then finds that the ratio rose substantially by the mid-1970s so that US$29 billion was in circulation beyond the figure required for legitimate transactions, assuming that the illegitimate cash circulated at the same velocity as the legitimate transactions. Consequently, informal employment was argued to represent more than 10 per cent of the officially calculated national income and the amount of currency in circulation was asserted to be equivalent to US$1,522.72 for a family of four. Such findings proved very influential with politicians and the media alike in bringing to the fore the notion that a sizeable amount of informal employment existed and this approach was subsequently adopted very widely (Cocco and Santos 1984, Matthews 1983, Matthews and Rastogi 1985, Meadows and Pihera 1981, Santos 1983, Tanzi 1980).[2]

However, this method suffers from such serious problems that it is inherently unsuitable for assessing either the extent of informal employment or the degree of informalisation in the advanced economies. First, cash is not always the medium for informal exchange. There is plenty of evidence that informal employment utilises cheques and credit cards as well (see below). Indeed, in some countries such as Italy, laws preclude the disclosure of information concerning bank accounts, so it is unnecessary to use only cash in informal employment (Contini 1982). Moreover, and as Smith (1985) identifies in the US, whether an informal payment is made in cash or by cheque depends on the same factors as determine the mode of payment in formal employment (i.e. the size of the transaction and the seller's confidence in the purchaser's cheque).

Second, this approach again has no way of distinguishing what share of the illegitimate cash circulation is due to informal employment and what proportion is due to crime, nor how it is changing over time.

Third, the choice of the cash–deposit ratio as a measure of informal employment is an arbitrary one that is not derived from economic theory (see, for example, Trundle 1982) and it is not clear why this was chosen rather than others.

Fourth, and again similar to the high-denomination notes approach, the cash–deposit ratio is influenced not only by the level of informal employment but by a myriad of other tendencies, often working in opposite directions to one another. As already stated, whilst methods of payment have changed, with credit cards and new interest-bearing assets reducing cash usage, increasing financial exclusion (e.g. refusal of credit cards and cheque accounts to the poor) resulting from the banks' 'flight' to affluent markets, has increased the use of cash for some populations. Mattera (1985) echoes these criticisms arguing that Gutmann fails to take account of factors other than those to do with informal employment, which might have contributed to the decline of the currency ratio. To try to take into account these factors, some commentators have refined the method used. Tanzi (1980), Matthews (1983) and Matthews and Rastogi (1985), rather than attributing the entire increase in the cash to current accounts ratio to greater levels of informal employment, instead focus upon only the proportion of the increase which can be shown to result from informal employment. These approaches, although more sophisticated, are not without their critics. Smith (1986) for instance, questions the appropriateness and value of the statistical variables employed by Matthews (1983), such as the identification of a positive causal relationship between rises in unemployment levels and the growth of informality.[3] Even using Matthews' own figures, informal employment expanded most rapidly at a time when unemployment rose comparatively little (Smith 1986). Thomas (1988), moreover, details some stark differences between the results obtained in Matthews (1983) and Matthews and Rastogi (1985), which neither study attempted to explain.

Fifth, the choice of a base period when informal employment supposedly did not exist is problematic, especially given that the results are very sensitive to which base year is chosen. O'Higgins (1981) shows in the UK that if 1974 is taken as the base year, 16.5 per cent of the currency in circulation was fuelling informal employment in 1978. However, if 1963 is taken as the base year, informal employment became negative in 1978. Thomas (1988) highlights the problems associated with the need to locate a year when informal employment did not exist. For him, the choice seems to be determined more by the availability of data than by any other factor and is somewhat *ad hoc*.

Indeed, work on the history of informal employment lends support to the idea that such work has been a feature of advanced economies throughout the twentieth century. Indeed, it could be argued that it has been in existence as long as there have been rules and regulations with regard to employment. Henry (1978), for example, cites examples of fiddling and tax evasion from the time of Aristotle, whilst Houghton (1979: 91) shows that in 1905, when tax was at a uniform rate of less than one shilling (5 pence) in the pound in the UK, a departmental committee reported that 'In the sphere in which self-assessment is still requisite, there is a substantial amount of fraud and evasion'. Smithies (1984), moreover, in a detailed case study of informal employment in five towns (Barnsley, Birkenhead, Brighton and Hove, Walsall and part of North London) between 1914 to 1970, clearly demonstrates a continuity in the prevalence of such activity. Any method which measures the size of informal employment based on the assumption that there was a time when it did not exist is thus founded on suspect grounds (see Henry 1978).

Sixth, to convert the estimates of informal cash into informal income, it is necessary to know the velocity of cash circulation in the informal sphere. No data exist on this, so the standard approach is to assume the same velocity as in the formal sphere. However, there is no evidence available to suggest why the two velocities should be the same (Frey and Weck 1983).

Seventh, it is impossible to determine how much of the currency of a country is held domestically and how much abroad. Some of the cash which the cash–deposit ratio assumes is held domestically will be doubtless held abroad causing an exaggeration in the estimate of informal employment.

In a bid to overcome one of the principal problems with this approach (i.e. that informal transactions are assumed to occur in cash), the next approach relaxes this assumption.

The monetary transactions approach

Recognising that cheques as well as cash are used in informal transactions, Feige (1979) measured the magnitude of informal employment by estimating the excess in total quantity of monetary transactions over the level that would be predicted in the absence of informal employment. As evidence that cheques as well as cash are used in informal transactions in the US, Feige (1990) quotes a study by the Internal Revenue Service (IRS) showing that between a quarter and a third of their estimate of unreported income was paid by cheque rather than currency. Many more studies have identified similar tendencies. In Norway, for instance, Isachsen *et al.* (1982) find that in 1980, about 20 per cent of informal services were paid for by cheque, whilst in Detroit, Smith (1985) provides a higher estimate in the realm of informal home repair, displaying that bills were settled roughly equally in cheques and cash.

By relaxing this cash-only assumption, the unsurprising result is that monetary transactions approaches generally produce much higher estimates of the size of informal employment than the previous approaches discussed. For instance, Feige (1990) reports that the US underground economy as a proportion of total reported adjusted gross income (AGI) rose from 0 in 1940 (the base year) to 20 per cent in 1945, declined subsequently to about 6 per cent in 1960, increased rapidly to reach 24 per cent in 1983 and then declined again to about 18 per cent in 1986.

In general, however, and before accepting such findings, it must be recognised that this approach has suffered from exactly the same problems as the cash–deposit approach discussed above. The only problem overcome is that it has accepted that cheques can be used in informal transactions. Here, therefore, these criticisms will not be repeated. Instead, and to conclude this review of the indirect monetary methods, the essential point to recognise is that the inherent problems with all of these methods for evaluating the volume of informal employment in advanced economies and whether informalisation is occurring raise grave doubts about the validity of their findings.

Indeed, it is on the whole only those who have been heavily involved in developing one or other of these indirect monetary methods who still attempt to defend them. The vast majority of commentators on informal employment resoundingly reject their usage as an accurate measure of the extent and character of informal employment. Smith (1986: 106), for example, concludes that 'Estimates of the size of the black economy based on cash indicators are best ignored', whilst Thomas (1988: 180) finds that 'the methodology underlying the monetary approaches . . . rests upon questionable and generally untestable assumptions and . . . the estimates they have generated are of dubious validity'. In consequence, we now examine another method of assessing informal employment that again uses monetary methods but in a more direct manner.

The income/expenditure discrepancies method

This approach seeks to measure the volume of informal employment by examining the difference in the estimates of expenditure and income either at the aggregate national level or through detailed micro-economic studies of different types of individuals or households. Again assuming that even if those involved in informal employment are able to conceal their incomes, they will not be able to conceal their expenditures, it is thought that one can identify the extent of informal employment by comparing income and expenditure levels.

Aggregate level studies, therefore, take the discrepancy between national expenditure and national income as an estimate of the size of informal employment. In the US, for example, Paglin (1994) examines the discrepancy between household expenditure and income surveys published annually in the Bureau of Labor Statistics Consumer Expenditure Survey (CES). He finds that between 1984 and 1992, informal employment has declined from 12.4 per cent of personal income in 1984 to 9.6 per cent in 1992, or from 10.2 per cent to 8.1 per cent of GDP over this period. This, he asserts, is principally due to the growth of formal employment during the 1980s in the US. Nevertheless, he finds that in 1992, 10.2 per cent of households were income-poor but consumption-rich and asserts that this is due to the existence of informal employment. In a bid to identify the types of household engaged in such employment, Paglin (1994) asserts that the poorest 20 per cent of households had an average after-tax income of US$5,648 in 1991 but an average expenditure level of US$13,464. He then takes a major logical leap to conclude that a sizeable number of the income-poor households are engaged in informal employment, failing to consider whether this could be due to other factors (e.g. retirement household spending, households between jobs, major one-off expenditures on costly items).

Other studies that use income/expenditure discrepancies to shed light on informal employment start at the disaggregated rather than the aggregated level. In the UK, these have used the Family Expenditure Survey (FES) as their data source (see Dilnot and Morris 1981, Macafee 1980, O'Higgins 1981). Comparing households' income and expenditure in 1,000 out of the 7,200 households surveyed for the 1977 FES so as to examine whether some households appear to live beyond their means, Dilnot and Morris (1981) employ a variety of 'traps' to exclude from their calculation such discrepancies as might be explained by factors other than informal employment (e.g. high expenditure due to an unusual major purchase or to the running down of accumulated wealth). After all adjustments, Dilnot and Morris (1981), assuming that tax evasion existed in any household whose expenditure exceeded its reported income by more than 15 per cent, derived upper and lower estimates of its extent. They reveal that 9.6–14.8 per cent of households evaded taxes and that tax evasion was equivalent to 2.3–3.0 per cent of the GNP in 1977. Such evasion, moreover, was found to be more prevalent amongst the self-employed (who understate their income by between 10–15 per cent) and part-time employees than those in full-time employment. Smith (1986) has further reinforced this in a study of the 1982 FES where he concludes that the self-employed understate their income by between 10 and 20 per cent.

O'Higgins (1981), however, has cast doubts on the accuracy of such data. He suggests that it could be an underestimate because 30 per cent of households refuse to participate in the FES, and it is possible and plausible that a greater proportion of non-respondents participate in informal employment than the 9.6 per cent of respondents suggested by Dilnot and Morris' (1981) lower bound estimate and probably to a greater extent than the average weekly figure of £31 they identified. As O'Higgins (1981) argues, even if as few as 25 per cent of non-respondents engage in informal employment to the extent of £31 weekly, the lower bound estimate would be raised by almost half, yielding an adjusted lower estimate of 3.5 per cent of GNP.

There is little doubt that this method has advantages over the more indirect monetary methods considered above, not least due to the fact that it relies on relatively direct and statistically representative survey data. The disadvantages, however, are manifold (see Thomas 1988, 1992, Smith 1986). For the discrepancy to represent a reasonable measure of the level of informal employment, a number of assumptions need to be made about the accuracy of the income and expenditure data. Take, for example, the expenditure side of the equation. Estimates such as those made by Dilnot and Morris (1981) are dependent upon the accurate declaration of expenditure to government interviewers by respondents. Mattera (1985) suggests that it is somewhat naive to assume that this is the case. Equally convincing is the criticism that, for most people, spending is either over- or underestimated during a survey in so far as records of spending are kept by few members of the population, compared with income, which for employees comes in regular recorded uniform instalments. Moreover, at an aggregate level at least, such expenditure will omit informal purchases by both formal and informal incomes (Dallago 1991). Those studies conducted of expenditure on a household level also suffer from the fatal flaw of only examining final demand (i.e. consumer expenditure), not intermediate demand (i.e. business expenditure) for informal goods and services. As such, it ignores informally produced intermediate demand, such as informal sub-contracting as well as off-the-books employment by formal enterprises (Portes 1994). On the income side, meanwhile, it is not possible to know whether the income derives from criminal or informal activities, or even whether it derives from wealth accumulated earlier such as money savings. In addition, and so far as studies such as the FES are concerned, there are problems of non-response as well as underreporting (Thomas 1992).

Consequently, it may be difficult to gain accurate information on the extent of informal employment from this approach. Weck-Hanneman and Frey (1985) in Switzerland bring such problems of the validity of income and expenditure data to the fore when they report that the national income tends to be larger than expenditure. Consequently, according to this method, Swiss informal employment is negative. This is nonsensical and reveals that the discrepancy does not display the level of informal employment but is due to other factors. As Frey and Weck (1983: 24) conclude about these monetary methods 'One of the main shortcomings of all these approaches is that they do not concentrate on the causes and circumstances in which a shadow economy arises and exists.' Nor do they explore the character of informal employment beyond crude estimates of its sectoral or occupational concentrations. To answer such issues, it is the more direct approaches to informal employment which need to be examined.

More direct survey methods

To survey directly the magnitude and character of informal employment, one can investigate suppliers and/or purchasers of informal goods and services with regard either to the volume or the value of the exchanges. To examine the *volume* of informal employment, therefore, studies can ask either households or in theory businesses whether they have used informal employment, formal employment or unpaid work to complete specific tasks and thus measure the relative importance of different forms of work. Alternatively, one can ask sellers of informal employment about their level of participation in specific activities. In practice, much of the research to assess the volume of informal employment has been conducted on a household level and requests information from respondents both as suppliers and purchasers of informally-produced goods and services (e.g. Leonard 1994, Pahl 1984, Warde 1990). To explore the *value* of the informal purchases and/or sales, meanwhile, the amount of money earned by sellers, or spent by consumers, with regard to informally produced goods and services can be examined. Again, most studies have tended to investigate respondents as both purchasers and sellers (e.g. Fortin *et al.* 1996, Isachsen *et al.* 1982, Lemieux *et al.* 1994), although some examine respondents only as purchasers (e.g. McCrohan *et al.* 1991, Smith 1985).

22

In each type of survey, moreover, and whether the value or volume of the goods and services purchased or supplied is examined, this data can be collected in either a quantitative manner such as through closed-ended questionnaires (e.g. Fortin *et al.* 1996, Isachsen *et al.* 1982) or more qualitatively through detailed ethnographic techniques (e.g. Howe 1988, 1990). On the whole, and this is perhaps a reflection of the lack of data on this subject, most studies have opted for primarily quantitative techniques and then frequently employed more qualitative methods in a secondary capacity for in-depth explanation of the findings (e.g. Leonard 1994, Pahl 1984). Indeed, even studies relying primarily on ethnography such as the study by Howe (1988) conduct some interviews as a quantitative basis for their ethnographic material.

Finally, although such direct studies could be carried out on either national, regional or local population samples, in most cases, they have focused upon particular localities (e.g. Barthe 1985, Leonard 1994, Pahl 1984), socio-economic groups such as homeworkers (e.g. Phizacklea and Wolkowitz 1995) or industrial sectors such as garment manufacturing (e.g. Lin 1995). Indeed, unless governments decide to invest in conducting such direct studies at a national level, it seems likely that these foci of investigation will remain in the future.

For those who employ the indirect approaches, the major criticism of these direct methods is that the researchers naively assume that people will reveal to them, or even know, the character and magnitude of informal employment in their lives. It is thought on the one hand, that purchasers of informal goods and services may not even know if it is being offered informally or formally and on the other hand, that sellers will be reticent about disclosing the nature and extent of their informal employment since it is illegal activity.

The former point is doubtlessly true. For example, if a purchaser has his/her external windows cleaned or purchases some goods from a market stall, s/he might assume that this money is not declared when this is not necessarily the case, or vice versa. In other words, although consumers may often assume that goods and services bought in certain contexts are informal whilst in other con-texts they are not, their assumptions are not always correct. Many goods acquired in formal retail outlets, for example, may actually have been produced and even sold informally (e.g. in illegally inhabited shop premises) without the knowledge of the consumer. Not all those paid in cash, meanwhile, are neces-sarily working informally, just as some of those accepting cheques may be tax evaders. On the whole, therefore, although people who purchase informal goods and services may be more willing to reveal that it is purchased informally, they cannot be sure whether this is indeed the case.

It is not necessarily the case, however, that those offering informal labour will be untruthful in their dealings with researchers. Indeed, such a criticism of the direct approaches has been refuted many times although it comes back again and again. Pahl (1984), in his study of the Isle of Sheppey, questioned people both as suppliers and purchasers. He found that when the results from individuals as suppliers and purchasers were compared, the same level of informal employment was discovered. The implication, therefore, is that individuals are not so secretive as many previously assumed about their informal employment. Just because it is activity hidden from or unregistered by the state for tax, social security and/or labour law purposes does not mean that people will hide it from each other or even from academic researchers. Similar conclusions have been drawn concerning the openness of research participants in Canada (Fortin *et al.* 1996) and the UK (Evason and Woods 1995, Leonard 1994, MacDonald 1994). As MacDonald (1994) reveals in his study of informal employment amongst the unemployed, 'fiddly work' was not a provocative subject from their perspective. They happily talked about it in the same breath as discussing, for instance, their experiences of starting up in self-employment or of voluntary work. This willingness of people to talk about their informal employment was also identified by Leonard (1994) in Belfast.

Perhaps a more salient criticism, especially of the approaches which measure the participation rates in informal employment according to different tasks undertaken by households (e.g. Pahl 1984, Warde 1990), is that they frequently only investigate the informal employment used to meet final demand (spending by consumers on goods and services), not intermediate demand (spending by businesses). In advanced economies such as the UK, the problem is that such final demand accounts for just two-thirds of total spending in the economy (HM Treasury 1994). Consequently, unless such techniques are combined with other methods that attempt to estimate intermediate demand, one will miss the informal employment that takes place in the other third of the economy. This is perhaps a valid criticism of these studies, but ignores the fact that their aim is not to measure the size of informal employment but, rather, how households are coping.

As Harding and Jenkins (1989) thus conclude about these direct methods, their particular strength is that they are actually designed to generate data on informal employment rather than make sense of data collected for other purposes. Consequently, they can be tailored to meet the need of the particular research problem. They are also much better tailored to explore the character of informal employment such as its distribution by gender, employment status, income and ethnicity than the indirect approaches discussed above. Of perhaps even greater salience is the fact that they can explore the motivations for, and causes of engagement in informal employment as well as the opportunities and barriers to participation. In consequence, these direct methods and their results will be returned to in much greater depth in the forthcoming chapters where attempts are made to explore both the magnitude and character of informal employment in the advanced economies.

Conclusions

In order to lay the foundations for understanding informal employment in the advanced economies, this chapter has evaluated the contrasting approaches commissioned to measure the magnitude and character of such activity. This has revealed that no one approach is perfect. In consequence, when trying to arrive at an overall picture of informal employment in the advanced economies, the impact on their findings of the methodology underpinning each study must be taken into account. If this is not done, then any review of the literature on informal employment will result in confusion. It is only by unpacking the methods used in each study that clarity and understanding of this phenomenon can be achieved.

3

THEORISING INFORMAL
EMPLOYMENT

Introduction

Having defined informal employment and reviewed the methods for examining such work, the aim of this chapter is to evaluate critically the emerging orthodoxy that the advanced economies are witnessing the growth of informal employment as it becomes an increasingly important survival strategy for marginalised populations (e.g. the unemployed, ethnic minorities and immigrants). Synthesising the findings of the vast array of direct surveys of informal employment undertaken in the advanced economies, first, the extent to which informal employment is growing or declining is analysed and, second, the character of this work is discussed. On the one hand, this reveals that the trajectory of economic development is not universally following either the path of formalisation or informalisation but is much more diversified than has so far been considered. On the other hand, the chapter shows that informal employment is not simply a survival strategy for the marginalised but rather, there is a heterogeneous informal labour market with a hierarchy of its own which reproduces the socio-spatial divisions prevalent in the formal labour market. To explain the heterogeneous nature of local informal labour markets, the final section demonstrates how their configuration in any particular area is the product of a 'cocktail' of factors. This cocktail is composed of a range of economic, social, institutional and environmental conditions which combine in diverse ways in different places to produce specific local outcomes. This sets the scene for the following chapters that examine in greater depth the social and spatial configuration of informal employment in the advanced economies and beyond.

The development path of advanced economies: formalisation, informalisation or heterogeneous development?

The fact that there is a substantial amount of economic activity taking place in the advanced economies which does not appear in the official statistics is nowhere disputed. However, the precise amount of this work and whether it is increasing or decreasing faster than measured economic activity is a matter of heated debate. Here, we evaluate critically both the belief that the latter half of the twentieth century has been characterised by an all-pervasive formalisation of economic life in the advanced economies, as well as the theory that in the last twenty years, the advanced economies, under pressure from globalised markets, are undergoing a process of informalisation.

The formalisation thesis

One of the most widely-held beliefs about economic development, and one which is seldom questioned, is that as economies become more 'advanced', there is a natural and inevitable shift of economic activity from the informal to the formal sphere (which we shall call 'the formalisation thesis'). Indeed, the formalisation of economic and social life is often the 'measuring rod' which defines Third World countries as 'developing' and the first world as 'advanced'. In this view, the existence of supposedly 'traditional' informal activities is a manifestation of 'backwardness' and it is assumed that they will disappear as economic 'advancement' and 'modernisation' occurs (e.g. Rostow 1960). However, there are good reasons why this view of a uni-dimensional trajectory of economic development has come under attack. Consequently, we now briefly examine the development path first of the advanced economies and second, of the Third World economies.

In advanced economies, there is little doubt that over the long wave of history, a relative formalisation of work and welfare has taken place. As Harding and Jenkins (1989: 15) assert,

> history may be viewed as the progressive encroachment of formality upon widening areas of social life, as a consequence of literacy and the introduction of ever more sophisticated information technology on the one hand, and the increasing power and bureaucratisation of the state, on the other.

With the creation of the modern nation-state, many aspects of informal organisations have been co-opted and formalised and there has been a gradual undermining of the various means of subsistence outside formal activity (Leonard 1994, Pahl 1984). Indeed, people have come to define themselves through their employment status (Kumar 1978). This has not meant, however, that the provision of informal work and welfare has ceased to exist, even if it has been out of sight for a long time. One has only to look at the continuously high level of unpaid self-provisioning (Gershuny 1992, Gershuny and Jones 1987, Gershuny et al. 1994) and the dominance of informal welfare provision in activities such as child-rearing (e.g. Windebank 1996) and elder-care (e.g. OPCS 1992) to see that there remains a sizeable element of economic and welfare activity in the advanced economies which has not been formalised. In sum, therefore, even if work and welfare have become formalised, particularly during the second half of the twentieth century, informal activity has not been fully eradicated.[1] Formal activity has nowhere become the sole means of meeting human needs, even in those command economies which sought to achieve such a goal (Korbonski 1981, Sik 1994). Instead, and if any tendency is to be discerned in recent economic history, it is quite the opposite.

As chapter 9 will reveal in greater detail, the supposedly natural culmination of formalisation – full employment and a comprehensive welfare state – can no longer be accepted as the end-state of economic development. Full employment has never really been achieved, except for at most thirty years or so following World War II in a few advanced economies and even then, it was only full employment for men, not women (Pahl 1984). Subsequently, however, the trend has been away from this apparently natural and inevitable end-state. Take, for example, the European Union (EU). Between 1965 and 1995, the share of the working-age population with a job fell from 65.2 per cent to 60.4 per cent (European Commission 1996b), revealing the increasing proportion of adults of working age without employment. Such statistics on levels of employment and non-employment, moreover, mask the extent to which under-employment is also rising as permanent full-time jobs are steadily being replaced with temporary and part-time employment (Nicaise 1996, Thomas and Smith 1995). In 1993, for example, almost one in three employees in the EU worked part-time and the 'life-span' of many jobs (even those defined as 'permanent') is rapidly declining. As the European Commission (1996a) report, the average 'life' of a job is 4–5 years in the EU, which is now similar to the USA. Consequently, despite the long-wave formalisation of the advanced economies, the hypothetical end-state of a stable full employment society has not only never existed but seems to be receding ever further into the past.

So too is a comprehensive welfare 'safety net' for those who find themselves excluded from employment. Taking the EU as an example, since it is within this group of nation-states that the most ardent supporters of high levels of welfare protection are to be found, the overarching message is that there is little room for optimism regarding the future development of formal welfare provision for the unemployed and underemployed. In all EU nations, welfare spending is under pressure as countries attempt not only to balance their books in preparation for the single currency, but also to reduce costs to compete with less-protected economies. As the European Commission (1995b) reports, growth in expenditure on social protection benefits per capita fell or remained static in six out of fifteen Member States during the 1990s compared with the late 1980s.[2]

It is not only in the advanced economies, however, that formalisation appears to be neither natural nor inevitable. Across the Third World, and as chapter 8 will reveal, formal employment is by no means universally growing. Although formal jobs have risen at a rate well in excess of the increase in the size of the labour force in many East and South-East Asian nations, there are whole swathes of the Third World not only at a very low base level so far as formal jobs are concerned but which are either standing still or are undergoing a process of informalisation. In low-income nations, for example, regular waged or salaried employment accounts for only 5–10 per cent of total employment and in many regions such as Latin America, sub-Saharan Africa, North Africa and the Middle East, employment growth has been either static or decreasing relative to the growth in the labour force (International Labour Organisation 1996).

Therefore, the formalisation thesis, upon which many of the current assumptions about the futures both of work and welfare are founded, must be treated with care. At a general level, formalisation has been a defining characteristic of twentieth-century social and economic life. That said, as the main expressions of formalisation, namely full-time stable employment and the welfare state, come under pressure, it is being increasingly recognised that nowhere did these become the sole means of the production of wealth and well-being. Indeed, some assume that if we are witnessing the exclusion of an increasing proportion of the citizens of advanced nations from both employment and formal welfare provision, then they must be turning towards informal activity as a means of survival. Therefore, can an informalisation of the advanced economies be discerned?

The informalisation thesis

The informalisation thesis describes a process by which the advanced economies are witnessing a growth of informal economic activity. Based on the assumption that informal employment is either a new form of advanced capitalist exploitation or a response to overregulation by the market, this increasingly popular view, held by right- and left-wing commentators alike, asserts that as the advanced economies fall into economic crisis and the neo-liberal project of deregulation takes hold, we are witnessing the renewed expansion of this form of work (Amin 1996, Castells and Portes 1989, De Soto 1989, Frank 1996, Ybarra 1989). As Castells and Portes (1989: 13) assert, what is new about informal employment in the contemporary period is that it is growing, even in highly institutionalised economies, at the expense of already formalised work relationships. Therefore, it is a new form of advanced capitalism rather than 'a mere "lag" from traditional relationships of production'.

Examining the evidence usually mustered to support this view, the principal problem is that it derives from the indirect macro-economic methods which, as the last chapter showed, are fatally flawed in their design, rendering the results of dubious quality and validity. For the same country in the same year, it is normal to find estimates of paid informal activity ranging from as little as 2–3 per cent to 30–35 per cent of GNP (see Barthelemy 1991). As Thomas (1988) displays, the extent of informal employment has been estimated to range from 1.5–27 per cent of GDP for the US and 2–22 per cent of GDP for the UK. In consequence, to rely on any of these indirect measures as evidence for the informalisation of the advanced economies is to close one's eyes to the inherent problems with the methodologies on which they are based.

The only other source of potential evidence, the micro-social studies of informal employment, although more direct and accurate in their portrayals of such work, tend to be snapshots, usually in specific localities. It must be recognised that some of these studies have found evidence in specific contexts, often in southern Europe, of an informalisation of the kind envisaged by Castells and Portes (1989), namely, the substitution of informal employment for formal jobs, usually through the mechanism of contracting out functions formerly undertaken within a company (e.g. Benton 1990). However, this is not the same as providing evidence of a general informalisation of economic activity. Indeed, the only two longitudinal micro-social studies of informal employment carried out to date both refute the notion of an informalisation of the advanced economies. Mogensen *et al.* (1995) in Denmark find that the proportion of the population engaged in informal employment has remained level at 13–15 per cent throughout the period from 1980 to 1995, whilst Fortin *et al.* (1996) in Quebec reveal that between 1985 and 1993, informal employment as a percentage of GNP stayed constant at 0.65 per cent. Micro-social studies, therefore, provide little suggestion of a general trend of informalisation.[3]

For some, it might be assumed that the demise of formal work and welfare noted above is sufficient evidence to indicate an informalisation of the advanced economies. This, nevertheless, is based on the assumption that formal and informal work are substitutable (see Castells and Portes 1989, Duncan 1992, Gutmann 1978) and that the rise of one leads to the fall of the other. The problem with this assertion is that numerous micro-social studies of particular localities suggest that the relationship at any time of these two forms of activity is not one of substitution but rather, one of complementarity (Cappechi 1989, Cornuel and Duriez 1985, Leonard 1994, Morris 1995, Pahl 1984). Therefore, there is an iterative process in place whereby formal and informal employment shape and are shaped by each other. Given this, it is thus misleading to view deformalisation as an indication of informalisation in the advanced economies. Although this might be the case in some places at certain moments, many micro-social studies show that this is not universally the case. The relationship between formality and informality, therefore, appears to be far more complex than simply one of universal substitutability (or even complementarity), as will be shown later in this chapter.

Transcending informalisation/formalisation: the heterogeneous development paths approach

It is clear from the above discussions first, that although there has certainly been a general formalisation of work and welfare, or social and economic life, over the course of the twentieth century, nowhere has this produced a situation where all wealth and well-being is created within the formal sphere of employment and state-provided welfare. Informal employment, alongside the various types of unpaid work, has always had a role to play in contributing to living standards. Second, we have seen that just because formal employment and welfare is in crisis, it must not be assumed that informal employment is developing to fill the void, since this would be to believe that informal provision is universally a substitute for formal provision. The conclusion that we must draw, therefore, is that formal and informal employment have always existed side-by-side and that the relationship between the two is more complex than simply one of substitutability.

Given this state of affairs, we argue that in order to understand the phenomenon that is informal employment in the advanced economies at the end of the twentieth century, it is necessary to examine such work in its social and geographical context rather than to seek evidence of universal trends. Indeed, examining the findings of the vast array of direct micro-social studies mentioned above and in chapter 2, it quickly becomes clear that there are different processes of formalisation and informalisation in evidence in different places at varying times. Before explicating more fully this socially and spatially refined theorisation of the level of informal employment in the advanced economies, it is first necessary to also explore the nature of such work.

Characterising the nature of informal employment: peripheral exploitative labour or a segmented informal labour market?

Based on the assumption that marginalised groups engage in informal employment as a means of survival (the 'marginality thesis'), the resulting debate concerning how to explain the existence of informal employment has revolved around whether such work is a leftover of classical capitalism (i.e. the formalisation thesis) or a new form of advanced capitalism (i.e. the informalisation thesis). This section reveals that whichever position is adopted, both views are founded upon an erroneous premise.

Informal employment: peripheral labour for marginalised groups?

The 'marginality thesis' which views informal employment as a form of peripheral labour for marginalised groups (e.g. migrants, the unemployed and poor) has a long historical antecedent and can be traced back to the contemporary origins of the study of such activity in the work of Hart (1973) in Ghana. Since then, writers have taken it up in both the developing and advanced economies (e.g. Button 1984, Elkin and McLaren 1991, Lagos 1995, Maldonado 1995, Rosanvallon 1980). In this perspective, informal employment is simply seen as an exploitative peripheral form of labour that is at the

31

bottom of a hierarchy of flexible types of employment and is undertaken mostly by marginalised population groups as a survival strategy. However, such a view covers but one part of all informal employment and one section of the informal workforce.

Informal employment, that is, is not a homogeneous category (Fortin *et al.* 1996, Jensen *et al.* 1995, Laguerre 1994, Leonard 1994, Mingione 1991, Pahl 1984, Renooy 1990, Waldinger and Lapp 1993). It ranges from 'organised' informal employment undertaken by employees for a business that conducts some or all of its activity informally to more 'individual' forms of informality. These latter cover not only forms of informal employment conducted by the self-employed concealing a proportion, or indeed all, of their earnings, but also casual one-off jobs undertaken on a cash-in-hand basis, such as for a neighbour, friend or relative. Not all informal employment, therefore, is low-paid, exploitative and undertaken by marginal groups. Of course, forms of organised exploitative informal employment exist, such as in labour-intensive small firms with low levels of capitalisation, utilising old technology and producing cheap products and services for local markets and export, which involve on the whole marginal populations engaging in low-paid exploitative activity (Lin 1995, Sassen 1991). However, there are also organised forms of informal employment which are autonomous in orientation, in highly-capitalised small firms which are modern and use high-technology equipment to produce higher-priced goods and services and whose informal employees are well-paid, employ higher skills and have more autonomy and control over their work, with relations between employers and employees based more upon co-operation than domination (Benton 1990, Cappechi 1989, Warren 1994). There are also forms of individual informal employment which are both deeply embedded in social life and have more in common with unpaid community exchange than they do with exploitative informal or formal employment. So, despite much of the emphasis in the literature on exploitative organised informal employment, the reality is that this is but one, albeit important, part of the spectrum of activities which constitute this form of work. If informal employment is not solely an exploitative form of peripheral employment for marginalised groups, then how should it be conceptualised?

Retheorising informal employment as a segmented informal labour market

Based on a detailed review of the numerous direct, mostly one-off locality studies of informal employment in the advanced economies, we make the assertion that existing alongside the formal labour market is a heterogeneous informal labour market composed of very different groups of people engaged in widely varying types of informal employment for diverse and contrasting reasons and receiving varying rates of pay. This informal labour market, to adopt a simplistic dual labour market model, ranges from 'core' informal employment which is relatively well-paid, autonomous and non-routine and where the worker often benefits just as much from the work as the employer, to 'peripheral' informal employment which is poorly paid, exploitative and routine and where the employee does not benefit as much as the employer. Just as in the formal labour market, moreover, there exist those excluded from even the most exploitative peripheral informal employment.

Take, for example, Lozano's (1989) study of flea markets in Northern California. Interviewing fifty dealers, she disaggregates voluntary and involuntary entrants. Voluntary entry occurs when an individual decides to leave their job in order to earn income through self-employment; or when informal employment is conducted as a source of additional income beyond that necessary to cover normal living expenses and levels of indebtedness. One fifth of respondents fell into this category. Involuntary entry, meanwhile, takes place when a person loses a job, income from jobs, pensions or welfare payments is insufficient to cover normal living expenses and levels of indebtedness or a person leaves school to enter the formal labour market and is unable to find a job. The remaining 80 per cent of dealers experienced one of these situations prior to their involvement in the flea market. As Lozano (1989) asserts, there is an important difference in the motivations of those who enter the flea market voluntarily or involuntarily. The former explain their participation in terms of intense dissatisfaction with the routine and authority of the formal employment-place and engage in informal employment for reasons of personal autonomy and flexibility and to 'be my own boss'. The involuntary joiners, however, are very different. Although a quarter of them claim to enjoy the autonomy of this work, none give this as a principal reason for their participation. They conduct such work because they need the money and provided they have not been out of work for a long time, because their desire for a job is strong.

As can be seen from this case study, informal workers conduct such work for a variety of reasons and have a range of interests that are neither identical nor necessarily complementary. Consequently, this form of employment cannot be regarded solely as a response by people reeling from the universal and homogeneous process of the restructuring of capital. Such a deterministic model would be erroneous. Not only do participants have multifarious motivations for engaging in informal employment but such employment often has rates of pay which match or exceed their equivalents in formal employment (Carbonetto *et al.* 1987, Ferman *et al.* 1978, Fortin *et al.* 1996, Henry 1978, Lozano 1985, Portes *et al.* 1986, Roberts 1992). Put another way, people's behaviour is determined by structural factors, but there is also agency involved. They have options. This dialectical relationship between structure and agency is the essence of the process that structures social formations (Giddens 1984, Thrift 1996). Thus, informal employment is not simply a voluntaristic choice but neither is it solely a structurally determined form of employment.

Who, however, receives these high-income informal jobs? And who is involved in the lower-paid more exploitative informal employment? In other words, how is the informal labour market configured socially (e.g. by gender, ethnicity and employment status)? In order to answer such questions, we now take a case study to introduce some of the issues involved.

The segmentation of the informal labour market: a case study of Quebec City

In 1986, 2,134 adults aged 18 years or over were surveyed in the Census Metropolitan Area of Quebec City, with a 63.8 per cent response rate (Fortin and Frechette 1986, 1987, Lemieux *et al.* 1994). The main sample (1,878 people) was of the random-cluster type. This was supplemented by a small quota sample (256 persons) to compensate for difficulties in reaching people in some areas and in some socio-economic groups. Defining informal employment as income not reported in the income tax statement, this could in theory have included both informal employment and wider criminal activity. However, in practice, they find that criminal activities were rarely reported in the survey. As Table 3.1 reveals, although 8.5 per cent of the total sample report working in informal employment for a mean of 357 hours annually yielding an average income of C$2,006, participation in this form of employment is not evenly distributed across all social groups. Men, students, the unemployed and lower-income populations were all more likely to conduct such employment. Super-ficially, therefore, these results could be taken to reinforce the marginality thesis that informal employment is an exploitative form of work mostly under-taken by those excluded from formal employment. However, closer analysis reveals that such a view does not fully describe the informal labour market.

For example, although the unemployed are more likely than the employed to engage in informal employment, they do not constitute the vast bulk of the informal workforce. Indeed, just 12.8 per cent of all informal workers are unemployed. The formally employed, meanwhile, constitute 35.2 per cent of all informal workers and earn more per hour for working informally than their more marginalised counterparts. So too do those who earn a higher formal income. Whilst those earning C$0–10,000 p.a. earn an average of C$4.96 per hour from their informal employment, the hourly wage rate rapidly rises with formal income so that those earning C$30–40,000 p.a. from formal employ-ment command informal wage rates of C$24.67 per hour. Informal employment, therefore, is clearly segmented along the lines of formal income and employ-ment status. It is a similar story when gender disparities are examined. Not only do a slightly higher proportion of men than women participate in informal employment (9.9 per cent compared with 7.1 per cent) but men receive higher hourly informal wage rates than women (C$6.93 compared with C$4.16 in 1986). The gender inequalities prevalent in formal employment therefore, are replicated in informal employment, displaying the gender segmentation of the informal labour market.

Table 3.1 Participation in informal employment in Quebec City, 1986

Characteristics	% of sample	% doing informal work	% of all informal workers	Hours of informal workers	Average informal earnings of informal workers	Average informal earnings of all population	Average $/hr	% of all informal work by: Hours	% of all informal work by: Value
All	100.0	8.5	100.0	357	2006	171	5.61	100.0	100.0
Gender:									
Men	48.7	9.9	56.7	331	2294	227	6.93	52.6	64.9
Women	51.3	7.1	43.3	391	1628	116	4.16	47.4	35.1
Employment status:									
Student	11.4	28.2	38.0	332	1976	557	5.45	35.6	37.4
Retired	5.1	1.9	1.2	120	490	9	4.08	0.4	0.3
Housekeeper	17.6	6.2	12.8	581	2251	140	3.87	21.1	14.4
Unemployed	4.0	27.4	12.8	369	1904	522	5.16	13.4	12.2
Employed	61.9	4.8	35.2	297	2034	98	6.85	29.5	35.7
Official income (C$):									
0–10,000	51.4	12.9	74.7	400	1984	256	4.96	84.3	73.9
10–20,000	17.2	7.0	13.4	286	2302	160	8.05	10.8	15.4
20–30,000	10.5	3.9	7.0	190	1943	75	10.23	3.7	6.8
30–40,000	9.9	2.0	2.2	58	1431	29	24.67	0.4	1.5
40,000+	11.0	2.3	2.7	104	1790	41	17.21	0.8	2.4

Source: Derived from Lemieux *et al.* (1994: Table 1).

Contrary to popular prejudice, furthermore, informal employment is not necessarily lower-paid than formal employment. Lemieux *et al.* (1994) find that although the average gross wage per hour is 13 per cent higher for formal employment than for informal employment (C\$7.98 compared with C\$6.99 in 1986), the average net wage per hour in formal jobs, which takes account of marginal tax rates and tax-back rates associated with social transfers, is lower than the average wage in informal employment. This is important, for it shows that informal employment is not merely an exploitative low-paid sector of the economy. It has a higher average net wage rate than formal employment. Nevertheless, the distribution of wage rates is wider than in formal employment. The standard deviation of the informal wage rate is larger than the standard deviation of the formal wage rate, suggesting that wage differentials are greater in this sector. This adds weight not only to the notion of a segmented informal labour market but also to the notion that it is perhaps segmented more intensely than the formal labour market. Indeed, whilst hours and wages are positively correlated in formal employment, they are negatively correlated in informal employment (Lemieux *et al.* 1994), indicating that there are many more in informal than formal employment who work long hours for little pay and short hours for high pay.

The important issue arising out of this case study is that the informal labour market is not solely composed of peripheral employment for marginalised populations. Instead, there is a segmented informal labour market that appears to a major extent to be segmented along similar lines to the formal labour market. However, some critical questions remain. Are there similar patterns of informal labour market segmentation in other localities, regions and nations? Do the unemployed, for instance, always constitute a smaller proportion of informal workers than the employed? Do they everywhere earn less than the employed? Does gender always relate in the same way to the informal labour market? What is the role of ethnicity? In sum, does geography matter? Part II of this book will therefore examine the relationship of employment status, gender and ethnicity to informal employment by reviewing a wide range of studies similar to the above which have been conducted throughout the advanced economies. Here, however, and before this is analysed, we need to provide a theorisation to explain the heterogeneous and segmented nature of informal labour markets in the advanced economies.

Explaining the magnitude and character of informal employment: the socio-spatial divisions of informal labour thesis

To explain why the growth, magnitude and character of informal employment varies socially and spatially, we argue that its configuration in any place derives from how economic, social, institutional and environmental conditions combine in multifarious 'cocktails' in different places to produce specific local outcomes. The consequence is that at any time, there is neither universal formalisation nor informalisation across the advanced economies but, rather, different processes in varying places. Neither is there one configuration of the informal labour market which is homogenous across all places. Instead, different configurations prevail in different areas. In this section, therefore, we move to outlining the various conditions that regulate the configuration of informal employment in any locality.

Before commencing this review of how economic, social, institutional and environmental conditions produce particular socio-spatial configurations of informal employment, it is important to make one point very clear. Many previous analyses of informal employment have tended to emphasise one condition or set of conditions over others when seeking to understand the level and/or nature of informal employment. First, for example, there are the economically deterministic approaches arguing that it is primarily economic circumstances such as increasing labour costs, competition from cheaper foreign goods or sub-contracting work arrangements which are leading to the growth of informal employment (e.g. Portes and Sassen-Koob 1987). Second, there are primarily institutionally focused approaches arguing that specific government regulations (e.g. tax laws) shape the extent and nature of informalisation (e.g. Warren 1994, Weiss 1987). Third, there are culturally-orientated approaches explaining informalisation in terms of factors like social networks (e.g. Howe 1990, Morris 1987) and finally, environmentally-deterministic approaches seeing inner-city, urban or rural areas as somehow inherently more conducive to such a form of work. Here, however, the argument is that there is a need to examine all of these factors and how they combine to produce specific local outcomes. This is because it is not simply the case that the mere existence of a particular condition (e.g. high taxes or lax enforcement of benefit laws) necessarily leads to informalisation or a particular configuration of the informal labour market. Although there are general tendencies, it is the ways in which a condition combines with other factors to produce specific local outcomes which is more important that the mere existence of a single condition in itself.

Economic regulators

Many explanations of the magnitude and character of informal employment rely heavily on economic explanations (Benton 1990, Portes and Sassen-Koob 1987) when explaining informalisation and the nature of the local informal labour market. Here, we outline the key regulatory conditions, which we derive from previous studies of informal employment, that are influential in configuring such work.

Level of affluence and employment

Many have demonstrated the existence of a strong correlation between the magnitude of informal employment and the level of affluence and formal employment in a locality. The problem, however, is that some have found that the higher the level of affluence and employment in a locality or region, the greater is the amount of informal employment, whilst others have discovered the opposite (e.g. see chapter 4). It is clear, therefore, that affluence alone is not a good indicator of the extent of informal employment. Nevertheless, affluence appears to be a much better indicator of the character of informal employment with more affluent areas and social groups carrying out more autonomous and better-paid activities.

Industrial structure

The term 'industrial structure' refers to the size and diversity of enterprises in a given locality. Most studies find that relatively little informal employment takes place in areas dominated by a few large companies because the skills acquired in such industries are less transferable to individual informal employment and such large companies less often use organised informal work than smaller firms (Barthelemy 1991, Howe 1988, Van Geuns *et al.* 1987). Informal employment usually flourishes, meanwhile, in local economies with a plethora of small firms (Barthelemy 1991, Blair and Endres 1994, Capecchi 1989, Pahl 1987, 1990, Sassen 1996, Van Geuns *et al.* 1987, Witte 1987) since trade unions are less active and organised; there is greater scope for employing organised informal work either directly or through sub-contracting arrangements; and skills acquired are more likely to be useful in individual informal employment. Similarly, many have drawn a strong positive correlation between the level of self-employment and the extent of informal employment (O'Higgins 1981, Pahl 1990). As Mingione (1991) has revealed, however, this supposedly close connection is very controversial and needs further clarification. Put another way, informalisation is not always more prevalent in areas dominated by small firms and self-employment. Although tax evasion may be higher amongst the self-employed than the employed (e.g. O'Higgins 1981, Pahl 1990), it is but one part of all informal employment. There is also evasion by employees and companies. Likewise, although small firms may evade taxes to a greater extent than large companies, this is also just one part of informal employment. There is also individual informal employment conducted by individuals on a small-scale basis who are not even registered as self-employed. Thus, one cannot automatically read off from the level of self-employment or small firms in a locality that informal employment is necessarily higher. Indeed, the magnitude of informality amongst small firms and the self-employed varies considerably between nations, regions and localities due to a host of other factors, such as the structure of taxation and the level of affluence and employment.

Sub-contracting

In some localities, a link has been established between the flexible production process of sub-contracting and the growth of informal employment (Beneria and Roldan 1987, Benton 1990, Thomas and Thomas 1994, Williams and Thomas 1996). This process is seen to go hand-in-hand with a change in the industrial structure towards small firms. Workers previously employed in large firms have been forced to seek jobs in smaller marginal firms or to do piece-work in their own homes. In economically vulnerable areas, the result has been a weakening of the workers' bargaining position, decreased wages, reduced health and safety conditions and a loss of state benefits (Benton 1990, Thomas and Thomas 1994). Such a deterioration in working conditions, however, is not always the inevitable outcome of sub-contracting. In parts of Italy, the result is thriving enclaves of small informal enterprises that have adapted to changing market demands for specialised products while sustaining relatively high wages and good working conditions, such as in Emilia Romagna (Capecchi 1989). Similarly, Benton (1990) explains that in the electronics industry around Madrid, professionals and technicians who envisaged opportunities in specialised markets not controlled by large firms initiated numerous small and medium-sized firms. These new firms have tended to sub-contract the main phases of assembly and production to skilled worker-entrepreneurs. Therefore, flexibilisation through sub-contracting does not necessarily and inevitably lead to exploitative kinds of informal employment. It depends on the local economy and the local labour market conditions as well as the local political circumstances. Neither is it always a prerequisite for informalisation. Some areas lacking such sub-contracting arrangements, such as Grand Failly (Legrain 1982) have experienced high levels of informalisation due to their social and environmental context (i.e. deeply embedded social networks and a rural surrounding).

Tax and social contributions

There is vast body of literature in economics which examines the relationship between informal employment and tax levels (e.g. Gutmann 1977, Jung et al. 1994, O'Higgins 1981, Scott and Grasmick 1981, Tanzi 1980). Although some argue that this condition alone is sufficient to explain the growth of informal employment (Geeroms and Mont 1987, Klovland 1980, Matthews 1982), it is but one, albeit important, part of the 'cocktail' of explanatory factors which cannot be ruled out. Generally, as taxes rise in a nation or region, the differential cost of using formal rather than informal employment increases for both consumers and firms. The incentive to work informally also intensifies, as does the differential cost for the consumer of using formal rather than informal employment (Feige and McGee 1983, Frey and Weck 1983, Gutmann 1977, Houston 1990, Renooy 1990). However, not all individuals will avoid such burdens by employing or offering themselves as informal workers. Instead, some consumers will engage in self-provisioning. Neither do all companies automatically switch some or all production to informal modes. It depends on what other options are open to them in their locality.

Indeed, it is not just the level but also the structure of tax and social contributions which influences the configuration of informal employment. This shapes who participates in informal employment and the nature of the work. In general, in places where taxes and social contributions are raised more from companies than individuals, organised informal employment might well be more prevalent, whilst if these contributions are raised to a greater extent from individual workers, then individual informal employment might be more likely (see Barthelemy 1991). In the UK, for example, where non-wage labour costs (e.g. employers' social security contributions, pension and health insurance) are just £18 for every £100 in wages, compared with £44 in Italy, £41 in France, £34 in Spain and £32 in Germany (DfEE 1996), it might be expected that organised informal employment will be lower than elsewhere. This, however, is not always the case. The way in which the level and structure of taxation influences the configuration of informal employment is dependent upon a range of other factors such as attitudes towards tax evasion, labour and welfare laws, and the interpretation and enforcement of rules and regulations.

Social regulators

Besides economic regulators of the extent and nature of the informal labour market in an area, there are also a host of social regulators that shape, and are shaped by, the influence of these economic conditions.

Local and regional traditions, social norms and moralities

In some areas, informal employment is more acceptable than in others, for example because feelings of resentment and of being let down by the state lead to less acquiescence to its laws (Howe 1988, Kroft *et al.* 1989, Legrain 1982, Leonard 1994, Van Geuns *et al.* 1987, Weber 1989) or because informal employment is undertaken more for social than economic reasons (Barthelemy 1991, Cornuel and Duriez 1985, Komter 1996). As Barthelemy (1991) points out in France, in the Auvergne, Limousin and Poitou-Charente regions, which are basically rural in nature, informal employment is a way of life rather than a tax fraud, a 'paid' element in community exchange. There is, in other words, a cultural mediation of economic processes. This shapes not only the level but also the structure of informal employment. Traditions of married women's employment, distinctive strategies of getting by, forms of collective communal action and many other practices are all mediated by cultural norms and traditions which influence the configuration of the informal labour market.

However, although local cultural traditions and attitudes shape the level and structure of informal employment, this is not usually taken to imply that there is a different set of cultural values attached to people who work informally compared with those who engage in formal employment (e.g. Lysestol 1995, Renooy 1990). Nevertheless, moralities are place-specific. Frey and Weck (1983) for example, find considerable cross-national differences in tax morality (i.e. the willingness to cheat on taxes) and Chavdarova (1995) in Bulgaria identifies clear differences between localities with low tax morality being highest in the big cities.

The nature of social networks

Dense social networks have been identified as a major contributory factor in explaining the existence of high levels of informal employment (Chu 1992, Legrain 1982, Meert et al. 1997, Mingione 1991, Morris 1988, St Leger and Gillespie 1991, Van Geuns et al. 1987, Warde 1990). Leonard (1994: 3) concludes that 'stability of residence, duration of unemployment, social homogeneity, extensive kinship and friendship networks and low levels of access to external resources are all conducive to the development of informal economic practices'. In this view, there is a kind of social capital seen to exist in certain communities that aids those who lack the economic resources to get by.

Dense social networks alone, however, are insufficient to lead to informal employment if all their members are unable to pay for such goods and services. As Turner et al. (1985) find in their study of a former coal-mining village in Scotland, despite the presence of kinship networks (60 per cent of the population had kinship links with others in the area), in the long run, unemployment pushed individuals into a marginal, isolated and excluded position peripheral to mainstream community life. Lysestol (1995) finds the same in Norway. As Morris (1993) discovers in the UK, moreover, the long-term unemployed often only mix with other long-term unemployed and thus are less likely to be offered opportunities for engaging in informal employment. However, it must be remembered that she was studying an area characterised by an industrial structure undergoing contraction and dominated by large enterprises. Dense social networks in conjunction with other factors (e.g. in areas with a wide socio-economic mix or rural areas) can lead to informal employment alleviating poverty, if not providing a complete survival strategy (Jessen et al. 1987, Weber 1989).

Socio-economic mix

It has been frequently found that an area which combines a population with high income but little free time with a population who have low income but much free time has both a relatively high proportion of the total work load undertaken using informal employment and a higher amount of such work

41

overall taking place (Barthelemy 1991, De Klerk and Vijgen 1985, Pestieau 1984, Portes 1994, Renooy 1984, Terhorst and Van de Ven 1985). This, nevertheless, is not always the case (Meert *et al.* 1997). The existence of social apartheid whereby there is little overlap between the networks of these two distinct populations may prevent this arrangement. So too may strict enforcement of social security regulations coupled with a low tax regime since suppliers would feel restricted from participating and customers would receive lower marginal benefits from using informal rather than formal employment.

Education levels

The more educated the person, the more likely s/he is to supply informal labour and to perform autonomous well-paid informal employment whilst those with fewer qualifications engage in more exploitative poorly-paid informal employment (Bloeme and Van Geuns 1987, Del Boca and Forte 1982, Fortin *et al.* 1996, Ginsburgh *et al.* 1987, Lemieux *et al.* 1994, Pestieau 1985, Renooy 1990, Sassen-Koob 1984). Relatively deprived areas, which have lower qualification levels, will thus generally perform less informal employment and that which takes place will be of the more exploitative variety. Again, however, there will be other configurations depending on the way in which this human capital factor is mediated by other economic, institutional, social and environmental conditions in any area (e.g. the structure of tax and welfare benefits and the degree to which they are enforced).

Institutional regulators

Besides economic and social regulators of the level and character of informal employment, there are also a range of institutional conditions which influence such work.

Labour law

By definition, and as revealed in chapter 1, government rules and restrictions create informal employment. As labour law differs between nations, so too does the level and nature of informal employment. Take, for example, industrial homeworking (see Boris and Prugl 1996, Phizacklea and Wolkowitz 1995). In Spain before 1980, and in contrast to nations such as the Netherlands and Britain, industrial homeworking was illegal. Thus, all industrial homeworking was by definition informal employment. As a result, companies needing such work as a competitive strategy were forced into informal employment, as were workers, such as married women with young children, who could not participate in other forms of work due to their family circumstances. Although in 1980 the law changed so that homeworking again became legal, the tradition of conducting homeworking informally had become established and continued (Van Geuns *et al.* 1987). Furthermore, what might be legal nationally might not be locally. In the UK, for example, leasehold regulations attached to property in particular localities ban industrial homework in a range of activities that are legal according to national labour law. Such legal differences thus have an influence on the level of employment conducted informally. In some nations, attempts at flexibilisation of the workforce through sub-contracting to industrial homeworkers and extended or irregular working hours will have to resort to informal employment due to the nature of labour law. Labour law, in sum, defines the structure and character of informal employment. This is a general rule that cannot be refuted, although it does show that informal employment in itself is a fluid category that varies temporally and spatially.

Welfare benefit regulations

The unemployed in countries with poor access to permanent state benefits are more likely to engage in informal employment as a survival strategy (Del Boca and Forte 1982) because they have little to lose if caught, thus helping to explain the higher levels of informal employment in southern EU nations and the US. In contrast, those eligible to claim more generous benefits will have the major disincentive of losing them if caught undertaking informal employment, as Wenig (1990) has highlighted in Germany. Again, however, this general tendency will be mediated by other factors. Leonard (1994) in West Belfast, for example, shows that local cultural traditions, social networks and the lack of buoyancy in the local labour market as well as poor enforcement of regulations can lead to high levels of informal employment amongst the unemployed.

State interpretation and enforcement of regulations

In addition to actual differences in rules and regulations, the socio-spatial configuration of informal employment will also differ according to how these rules and regulations are variously interpreted and enforced across localities, regions and nations. Some authorities, both national and local, deliberately overlook informal employment so as to enable firms to compete in international markets and/or to help individuals and families raise adequate incomes, which would be difficult if 'regular' regulations were strictly enforced (Cappechi 1989, Lobo 1990a and b, Mingione 1990, Portes and Sassen-Koob 1987, Van Geuns *et al.* 1987, Warren 1994). Regulation is applied to informal employment, moreover, not only through passive tolerance of its existence but also through active practices to curtail or develop it (see chapter 9). For example, there is well-documented proof of support for informal co-operatives by local governments

(Portes and Sassen-Koob 1987, Vinay 1987, Warren 1994) and it can have an influence on the structure of the informal labour market, such as where economic development agencies encourage inward investors to use women home-workers on an informal basis (Gringeri 1996). However, on its own, this is again insufficient to determine the level and structure of informal employment.

Corporatist agreements

Labour is not only regulated by the state but by industry, company and individual agreements which can limit the rights of employees to take on outside jobs, do overtime and so on. Informal employment can be an attempt to circumvent these rules which thus become a factor in creating local and national variations in the level and character of informal employment. In the Netherlands, for instance, whilst most forms of industrial homeworking are legal, some are not. There are industrial sectors where the Collective Labour Agreements (CAOs) forbid homework (e.g. the printing industry), whereas there are others where regulations for homework have been drafted into such agreements (e.g. the garment industry). In France, there has been a statutory working week that can only be extended by formal mutual agreement between employers and employees. If such agreements have not been forthcoming, then one can assume that overtime has had to be carried out on an informal basis. Furthermore, it is not only the strength of the state that encourages some to resort to informal employment, but also the strength of unions. For example, Brusco (1986) argues that decentralisation and informalisation in Italy are responses to the preceding growth of union power and the constraints it imposes on large firms. Lobo (1990b) in Portugal asserts that the presence of strong and autonomous unionism in large enterprises and in proletarian regions has reinforced social guarantees and regulation in the areas in which they operate, but their lack of presence has increased deregulation and thus informal employment in weaker areas and small firms. As he asserts, it is only in this sense that one can assert that informalisation is a product of stronger regulation. He disagrees that it occurs for this reason in any other sense.

Environmental regulators

In addition to the above economic, social and institutional regulators of the configuration of informal employment, the final set of conditions which influence the magnitude and structure of local informal labour markets are environmental.

Size and type of settlement

Although little is known about how urban settlement size influences the magnitude and character of informal employment, rural areas have been shown to

44

undertake more informal work than urban areas (Hadjimichalis and Vaiou 1989, Jessen et al. 1987), whilst lower-middle-class and working-class urban areas are found to engage in more informal employment than more affluent urban areas (Ferman et al. 1978, Simon and Witte 1982). There are, however, many exceptions. Van Ours (1991), examining how households get small home repairs, car repairs and maintenance and ladies' hairdressing done, finds that rural households use less informal employment than households in small and large cities. Mogensen (1985), in Denmark, meanwhile, shows that the frequency of participation slowly decreases as one moves from large urban areas such as Copenhagen (17 per cent participation in informal employment) to the rural areas of Western Jutland (10 per cent participation). In Canada, furthermore, Fortin et al. (1996) find that the participation rate in informal employment is much lower in the rural area of Bas-du-Fleuve than in Quebec and Montreal. The implication, therefore, is that rurality alone is an insufficient determinant of the level of informal employment. It depends upon the presence or absence of a range of additional factors that have been discussed above.

Furthermore, and so far as the character of informal employment is concerned, it seems that rural areas are characterised more by autonomous forms of informal employment (Duncan 1992, Levitan and Feldman 1991, Meert et al. 1997) whilst organised informal employment seems to be more concentrated in urban areas (see chapter 6). Again, however, this depends upon a range of additional characteristics and how these combine to produce particular configurations of informal employment. It is not so simple that one can universally state that rural areas do more than urban areas or that more affluent urban areas do less than poorer urban areas (see chapter 7).

Type and availability of housing

Housing tenure seems to be an important condition influencing the quantity and quality of informal employment in a locality since the level and type of activity is heavily reliant on the tenure of a household. Areas with high levels of privately-owned households will undertake a wider range of informal employment than areas dominated by rented accommodation (Morris 1988, 1994, Pahl 1984, Renooy 1990, Windebank 1991) and are also more likely to use such employment to get a job done (Mogensen 1985, Renooy 1990). Furthermore, given that nearly half of all council tenant households of working age in the UK now have no earner (Schmitt and Wadsworth 1994), housing tenure is an important determinant of the spatial concentrations of no-earner households who have been widely identified as undertaking little informal employment (see chapter 4). In addition, the availability of a larger living space, particularly the availability of garden space, offers greater possibilities for participation in individual informal employment (Van Geuns et al. 1987).

Conclusions

In contrast to many of the assumptions which abound concerning informal employment, this chapter has shown that it is not a phenomenon extinguished by the modernisation of economic and social life in the second half of the twentieth century (the formalisation thesis) which has risen, albeit in a different form, from the ashes of economic restructuring since the 1980s (the informalisation thesis). Furthermore, it has been asserted that informal employment is not simply peripheral employment for marginalised populations. Instead, we have here argued that the magnitude and growth/decline of informal employment varies across localities, regions and nations and that rather than perceive such employment as a peripheral type of formal labour, it is more accurately portrayed as a segmented labour market with a hierarchy of its own. Drawing upon the vast array of direct studies of informal employment in particular localities, we have argued that the changing socio-spatial configuration of the level and nature of informal employment is the result of the way in which economic, social, institutional and environmental conditions combine in multifarious 'cocktails' in different places to produce specific local outcomes. The consequence is that at any time, there is not a process of universal informalisation across the advanced economies but, rather, different processes in varying places. Neither is there one configuration of the informal labour market which is the same across all places. Rather, there are different configurations in different areas.

Our intention in arguing for a recognition of the multifarious nature of informal employment, however, is not to compensate for the previously over-generalised and simplistic typologies concerning the magnitude and character of informal employment by concluding with an idiographic approach which descends into particularism. This would be to go too far to the other extreme, ignoring some important parallels and commonalities in the nature and extent of informal employment between localities, regions and nations both in the advanced economies and beyond. Instead, in Part II, the intention is to draw out the commonalities and explain them, whilst not losing sight of the variations, in how this informal labour market is configured both socially (e.g. by gender, ethnicity and employment status) and spatially. It is to this task that our attention now turns.

Part II

SOCIO-SPATIAL DIVISIONS IN INFORMAL EMPLOYMENT

4

EMPLOYMENT STATUS AND INFORMAL EMPLOYMENT

Introduction

The aim of this chapter is to evaluate critically the popular belief that the unemployed participate in and gain from informal employment disproportionately relative to the employed and thus, that such activity is a survival strategy of the economically excluded. Having gained currency throughout the 1970s and 1980s (Gutmann 1978, Isachsen and Strom 1985, Matthews 1983, Peterson 1982, Rosanvallon 1980, Simon and Witte 1982), this perception is particularly popular at present. This is manifested in media campaigns and public outcries about alleged cases of benefit fraud committed by 'welfare spongers' or 'scroungers' (see Cook 1989, 1997, Malone 1994, Pahl 1985b), as recently displayed by the 'shop a dole cheat' hotline in the UK (Brindle 1995). Politicians interested in reducing social security expenditure, moreover, have done little to assess the validity of this perception of the unemployed as 'villains' rather than 'victims' of labour market restructuring. This would be harmless if it did not have political ramifications. However, politicians have often exploited this view to legitimise reductions in welfare provision for the unemployed and to increase expenditure on policing the benefit system.[1] The frequent upshot has been more punitive policies towards the unemployed, such as the Jobseekers' Allowance in the UK.

Neither has such a picture of informal employees been questioned by much of the theoretical academic discourse. Contemporary analyses of the role of informal employment in capitalism, as the previous chapter revealed, widely assume that such work is simply another form of peripheral employment conducted by the marginalised and unemployed (e.g. Castells and Portes 1989, Sassen 1989). In France, for example, Rosanvallon (1980) asserted that high unemployment at the end of the 1970s did not result in the traditional French reaction of rioting in the streets because the unemployed, many of whom were not entitled to any benefits at all at the time, must be working informally. In the recession of the early 1980s in the UK, meanwhile, Parker (1982: 33) claimed that 'with high unemployment more and more people are getting caught up in the web of the underground economy', whilst in the heyday of

49

laissez-faire free marketeers, Robson (1988: 55) asserted that 'the informal economy is more feasible as an alternative prop to those who are out of work'. In the US, similarly, Stauffer (1995: 1) argues that 'the informal sector can act as an important buffer against unemployment', whilst Blair and Endres (1994: 288) assert that 'The role of the informal sector in providing a source of support for unemployed workers or individuals receiving public assistance is an important function of the unobserved sector'. The assumption underlying many analyses, therefore, is that 'a significant percentage of the officially unemployed are in reality working "off the books", being paid in cash without intercession of a tax collector' (Gutmann 1978: 26).

As Pahl (1988: 249) intimates, this belief concerning the relationship of the unemployed to informal employment 'is in danger of becoming a social scientists' folk myth'. Here, therefore, the studies of informal employment conducted throughout the advanced economies are examined to put this myth to the test of documented evidence. This analysis reveals that it is seldom the case that informal employment is concentrated amongst the unemployed and is thus a survival strategy of the economically excluded. Instead, it shows that the unemployed find it more difficult than the employed to augment their incomes through informal employment. The reasons for this configuration are then explained in terms of how various economic, social, institutional and environmental conditions frequently combine to produce a 'cocktail' which excludes the unemployed from such work. Nevertheless, we then show how cocktails of conditions can also be created in certain specific circumstances which enable other local informal labour market structures to be produced where the unemployed can and do participate in such work.

The extent of informal employment amongst the employed and unemployed

To what extent do the unemployed undertake informal employment? Are they more likely to engage in informal employment than the employed? Do they do a disproportionate share of such work? Take, for example, the studies conducted in northern EU nations. In the Netherlands, Van Geuns *et al.* (1987) find in all the six localities studied that the unemployed generally do not participate in such employment to the same extent as the employed. This is additionally found to be the case in the Netherlands by Van Eck and Kazemeier (1990) and Koopmans (1989). It is also echoed in studies carried out in France (Barthe 1988, Cornuel and Duriez 1985, Foudi *et al.* 1982, Tievant 1982), Germany (Glatzer and Berger 1988, Hellberger and Schwarze 1987) and in Britain (Economist Intelligence Unit 1982, Howe 1990, Morris 1994, Pahl 1984, Warde 1990). The overwhelming conclusion in northern EU nations, therefore, is that informal employment is primarily a means of accumulating advantage for those already in employment and that the number of regular 'working claimants' accounts for a very small proportion of all informal employees.

Similarly, locality studies in southern EU nations reveal that the unemployed claiming benefit are less likely to engage in informal employment than the employed. In Spain, the Ministry of the Economy estimates that 29 per cent of those in employment also have an informal job (in Hadjimichalis and Vaiou 1989), whilst Lobo (1990a) finds that just 12 per cent of those claiming benefit perform such work. Benton (1990) reveals that 65.7 per cent of all informal employees have a formal job, whilst just 5.2 per cent are working informally and receiving social security benefit at the same time. The remaining 29 per cent of informal workers, we must assume, are those unemployed not entitled to benefit. As Lobo (1990a) concludes, most informal employment is conducted by the employed or those unemployed and not claiming benefit rather than by the unemployed receiving benefit. In Greece, similarly, the Ministry of Planning recognises the fact that it is more likely to be the employed than the unemployed who have informal jobs when it states that 40 per cent of those employed in the private sector also have non-declared jobs as do 20 per cent of those in public sector jobs (in Hadjimichalis and Vaiou 1989). In Italy, meanwhile, the majority of studies again discover that it is the employed rather than the unemployed who constitute the vast bulk of the informal labour force (Cappechi 1989, Mingione 1991, Mingione and Morlicchio 1993, Warren 1994) and that their participation rates are widening over time (Mingione and Magatti 1995).

This finding is not only prevalent across Europe but also in North America (Fortin et al. 1996, Jensen et al. 1995, Lemieux et al. 1994, Lozano 1989). In a study of informal employment in the San Francisco bay area, Lozano (1989) finds that it is not the poor and unemployed who engage in informal employment but, rather, people who are driven by tensions on the shop and office floor, where frustration with managerial authority and bureaucratic control lead them to seek an escape through informal employment. Therefore, in this relatively wealthy region, it is a choice on the part of participants, the majority of whom are in high-income households. Informal employment is not only or necessarily the strategy of unemployed workers but is a voluntary choice on the part of some who are relatively well paid. Jensen et al. (1995) in rural Pennsylvania, moreover, find that it is more likely to be those in employment than the unemployed who engage in informal employment whilst Fortin et al. (1996) in Quebec provide evidence that the formally employed constitute a much larger proportion of the informal workforce than the unemployed.

There is a need, however, as identified in some of these studies, to distinguish between different categories of unemployed due to the significant variations in their participation in informal employment. How they disaggregate the unemployed, however, differs according to the nations under discussion. In northern EU nations, for example, the majority of studies distinguish between the long- and short-term unemployed. In Germany, for example, Hellberger and Schwarze (1987) find that whilst 16.7 per cent of the temporarily unemployed engage in informal employment, only 5.8 per cent of the permanently unemployed conduct such work. Morris (1995) comes to a similar conclusion in Britain, stating that this is perhaps because the short-term or temporarily unemployed retain many of the contacts, resources and skills which they gained from their employment. Engberson et al. (1993) come to much the same conclusion in the Netherlands.

Studies in southern EU nations, meanwhile, distinguish between the unemployed who are claiming benefits and those who are not when discussing participation in informal employment. The reason for this, as Reissert (1994) reveals, is that a much lower percentage of the total unemployed in southern EU states receive unemployment compensation benefits compared to northern EU nations. Examining the European Labour Force Survey, he finds that in 1990 the benefit recipient quotas (i.e. the number of benefit recipients as a proportion of the total unemployed) ranged from less than 20 per cent in southern EU nations such as Greece, Portugal and Italy to more than 80 per cent in Denmark and Belgium. Those not receiving social assistance are found to engage in informal employment to a greater extent than those who receive it (see, for example, Lobo 1990a and b). In the US, a similar distinction is again drawn resulting in the same finding (Blair and Endres 1994).

However, it is not only the extent of participation that is important to consider when examining the relationship between employment status and informal employment. There is also the issue of the nature or quality of the informal employment undertaken.

The character of informal employment undertaken by the unemployed and employed

Here, therefore, we shift our focus from an analysis of how participation in informal employment varies according to employment status and the variations in participation rates amongst different segments of the unemployed to the issue of the quality or character of informal employment undertaken by these groups. To do this, we consider the types of informal employment in which the employed and unemployed engage, their contrasting motivations for joining the informal labour force and their differing informal wage rates.

Types of informal employment undertaken by the employed and unemployed

The vast majority of studies find that the employed tend to engage in more autonomous, non-routine and rewarding informal jobs than the unemployed who conduct more routine, lower-paid, exploitative and monotonous informal employment (Fortin et al. 1996, Howe 1990, Lemieux et al. 1994, Lobo

1990a and b, MacDonald 1994, Pahl 1987, Williams and Windebank 1995a, Windebank and Williams 1997). A plumber employed by one of the large utilities, for example, may put in central heating for cash-in-hand during the weekend for a customer s/he has met during the course of his/her formal employment, whilst an unemployed person may only have access to low-paid homeworking activity such as assembling Christmas crackers which s/he has learnt about through an advertisement in a shop window or newspaper or through a friend who already carries out such an activity. The result, therefore, is a segmented informal labour market in which many of those who already have a formal occupation find relatively well-paid informal employment, often conducted on a self-employed basis, whilst the unemployed generally engage in relatively low-paid organised informal employment which tends to be more exploitative in nature.

Take, for example, Lobo's (1990b) analysis of the service sector in Portugal. He distinguishes between two types of informal worker. On the one hand, there are well-qualified self-employed informal workers, engaged in activities such as domestic repairs (e.g. washing machines, televisions and radios), who have the opportunity to undertake this kind of work because they also have another job, usually full-time. On the other hand, there are informal employees (e.g. in catering, retailing) who usually do not have a formal job and are female, young and poorly qualified. Many of these tend to work in very precarious jobs (Izquierdo et al. 1987) with low earnings (Rusega and De Blas 1985) or with very long hours when their income is compared with the employed (Lopez 1986). Lobo (1990a) thus concludes that there is a segmented informal labour market.

Given the multifarious nature of this informal labour market, it is apparent that not everybody engaging in informal employment will be forced into this employment by economic necessity and neither will all informal employees have low pay rates. Here, therefore, we first examine the different motivations for engaging in informal employment amongst the employed and unemployed and then analyse the variations in their pay rates for informal employment.

Motivations for engaging in informal employment

The most striking finding of studies examining why the unemployed and employed engage in informal employment are the major differences in the reasons given for their participation. For the employed, especially those with relatively high formal incomes, there is frequently an element of voluntarism involved in their decision. Indeed, the intentions for these informal workers are often as much social as economic in character (Cornuel and Duriez 1985, Felt and Sinclair 1992, Jensen et al. 1995, Komter 1996, Leonard 1994). Informality, therefore, is not solely a response to economic circumstances, at least for these workers. As Felt and Sinclair (1992: 60) discover in their research on the Great Northern Peninsula in Canada, 'the informal sector is more extensive than it need be, if getting by were the only motivation for participation'. This finding is repeated elsewhere. In relatively affluent commuter villages in France, for example, Cornuel and Duriez (1985) found that participation in informal employment was primarily undertaken for the reason of constructing and maintaining social networks. Renooy (1990) identifies similar motivations in the Netherlands. For many of the unemployed, however, their participation is more involuntary and undertaken for the economic purpose of earning sufficient income in order to get by (Howe 1990, Jordan et al. 1992, Leonard 1994, MacDonald 1994, Rowlingson et al. 1997). All of these studies show that where the unemployed engage in informal employment, financial need is the overwhelming motivation. For example, in a study of forty-five benefit fraudsters, Rowlingson et al. (1997) discover that their main motivation for working whilst claiming was to raise money to buy essential items or pay bills since they were finding it difficult to manage on benefit alone. This finding is echoed in all of the above studies.

Having contrasted the motivations of the employed and unemployed, however, several important points need to be made. On the one hand, although many employees engage in informal employment as much for social as economic reasons, very few of them see this as a substitute for formal employment so far as defining their social status and forging social relations are concerned (Jensen et al. 1995, Jessen et al. 1987, Lysestol 1995, Pahl 1990). Informal exchange is used simply as a means of consolidating these advantages (see Komter 1996). On the other hand, and so far as the unemployed are concerned, most undertake informal employment because a formal job is not available to them. Few see it as an adequate substitute for formal employment and the vast majority would forego this work if a formal job became available (Evason and Woods 1995, Jensen et al. 1995, Jessen et al. 1987, Lysestol 1995, MacDonald 1994). Put another way, there is not a counter-culture or sub-culture amongst those unemployed who participate in informal employment (Dean and Melrose 1996, Jessen et al. 1987, Lysestol 1995, Morris 1993). They possess similar aspirations to the majority of the population in that they want a formal job. Indeed, a study of ninety-six Dutch benefit recipients, of whom half had committed fraud, found that fraudsters had a stronger work ethic than non-fraudsters (Hessing et al. 1993).

Many unemployed people, moreover, like the employed, see little immoral in their participation in such work. In Quebec, just 9.4 per cent of informal workers saw it as immoral (Fortin et al. 1996). This can be explained by looking at studies elsewhere. In Dean and Melrose's (1996) study of welfare fiddlers in London and Luton, respondents distinguish between their own 'fiddling' which they felt to be harmless and more serious or organised forms of fraud that they did not. They thus impose their own moral limits or rules upon fiddling. This was also the finding of MacDonald (1994) in Cleveland in north-east England. None of the 214 unemployed respondents interviewed in 1990 and 1992 saw such work as a substitute for a formal job and where they engaged in such 'fiddly jobs', it was motivated by both economic necessity and to preserve self-respect. As such, they saw their own informal employment as morally justifiable and contrasted it with more serious cases of fraud that they did not. Indeed, according to Rowlingson *et al.* (1997), most claimants who break the rules do not think that they are committing a crime. Put another way, that which was short-term, motivated by family need and done for relatively little cash was acceptable. Fiddling as a way of life, however, was wrong. People who had an alternative to being on the dole (who could have a legitimate job due to their skills but used these to earn on the side whilst claiming) were condemned. This finding is echoed in the work of Jordan and Redley (1994) in south-western England. Henry (1978), who reports that the ability of people who fiddled to justify their activities was an important determinant of continued fiddling, also made such a distinction. Indeed, the same distinction is drawn by the benefit authorities who distinguish between 'serious' and 'trivial' fraud and concentrate on the former (MacDonald 1994).

Informal employment, therefore, is not solely a response to economic circumstances and indeed, when households are in financial crises, the most common response is to cut back on expenditure rather than to seek income-generating activities. In a study of Kirkcaldy, Main (1994) examines how respondents in financial crisis deal with it. Non-labour market responses are most frequently mentioned. For example, the most popular strategy is to cut back on luxuries (stated by 56.5 per cent of women and 58.5 per cent of men), followed by cutting back on necessities (41.9 per cent and 35.5 per cent), giving up holidays (23.3 per cent and 23.3 per cent), receiving financial help from family and/or friends (14.6 per cent and 13.0 per cent), working overtime (2.5 per cent and 11.8 per cent), taking a second job (2.7 per cent and 4.5 per cent), the partner taking a second job (2.5 per cent and 0.9 per cent) and going back to work (11.2 per cent and 2.7 per cent). The implication, therefore, is that labour market responses such as informal employment are not a primary coping strategy adopted by people when dealing with a financial crisis. Instead, their principal response is to reduce expenditure rather than try to generate more income. Consequently, even if the unemployed mostly engage in informal employment for primarily economic motivations, this does not make informal employment a principal coping strategy used by the unemployed. Any additional income received, moreover, is likely to come from kinship networks as loans or gifts rather than from informal employment (Foudi *et al.* 1982, Lysestol 1995, Morris 1993). Nevertheless, for those who do engage in such work and as shown, it is not seen as immoral. Rather, they differentiate between their own petty fiddling and that of others engaged in more serious crime.

Pay rates for informal employment: by employment status

Besides motivations, employed and unemployed informal workers can also be differentiated by their rates of pay. Indeed, it is the variation in informal wage rates which perhaps highlights better than any other variable the segmented nature of the informal labour market by employment status. As Van Eck and Kazemeier (1990) identify in the Netherlands, Fortin et al. (1996) in Canada, Mattera (1980) in Italy and Hellberger and Schwarze (1986) in Germany, there is a clear relationship between the level of informal earnings and employment status. As Table 4.1 reveals, in 1993, the employed in the Quebec region earned an average of C$10.66 per hour from informal employment whilst the unemployed earned C$7.94, that is, just 74 per cent of the level of the employed. Between 1985 and 1993, moreover, whilst the hourly informal wage rate of the unemployed increased by 13.3 per cent, it rose by 15.8 per cent for the employed, displaying the increasing polarisation of hourly wage rates between the employed and unemployed. At the same time, the percentage of all unemployed working informally decreased drastically from 24.5 per cent to 8.5 per cent (mostly due to the stricter enforcement of rules concerning working whilst claiming). The result is that the proportion of all informal earnings going to the employed rose from 41.4 per cent to 43.1 per cent between 1985 and 1993. Informal employment, therefore, increasingly became a vehicle for ameliorating the earnings of the employed in this region.

The finding that those in formal jobs earn more in informal employment than the unemployed is reinforced by Renooy's (1990) study of informal employment in the Netherlands. He finds that the average hourly pay reflects the skills required for the job and that since the unemployed tend to have lower or less marketable skills than the employed, they are unable to command high informal wage rates. If their skills are unwanted on the formal labour market, there is little reason to believe that they will be requested on the informal labour market. Instead, the type of work they engage in tends to be low-skilled and composed of activities where there is an abundant potential labour supply and thus low wages.

Table 4.1 Wage rates for informal employment in Quebec region: by employment status, 1993 and 1985

1993 (1985 in parentheses)	% doing informal work	% of all informal workers	Average wage/hr C$	% of informal work by: Hours	Value
All	4.8 (6.0)	100.0 (100.0)	7.66 (7.28)	100.0	100.0
Student	13.5 (19.4)	42.7 (31.2)	8.00 (6.65)	33.7 (31.5)	32.1 (27.4)
Retired	1.0 (1.0)	2.1 (0.9)	9.01 (N/A)	2.3 (–)	2.2 (0.2)
Housekeeper	4.5 (5.0)	9.3 (14.4)	5.06 (5.62)	14.3 (24.0)	9.4 (19.2)
Unemployed	8.5 (24.5)	11.5 (11.6)	7.94 (7.01)	13.2 (12.2)	13.2 (11.8)
Employed	2.9 (3.9)	34.4 (41.9)	10.66 (9.20)	36.5 (32.3)	43.1 (41.4)

Source: Derived from Fortin et al. (1996: Tables 3.1 and 4.2).

Their lower wage levels are also related to the way in which they find informal employment. As Van Eck and Kazemeier (1985) identify, informal employment found through employers or colleagues is by far the best paid, namely two or three times higher than informal employment found through other channels. Given that the unemployed have neither employers nor colleagues, the result is that they have limited access to these better-paid informal employment opportunities. Instead, their networks confine them to seeking work via family or acquaintances. The result is lower wage levels since the average wage for work conducted for one's family is Dfl 10, for acquaintances it is Dfl 14, for colleagues Dfl 20 and for employers Dfl 33 (Van Eck and Kazemeier 1985).

With regard to wages from informal employment, it is not simply the case that the unemployed receive lower informal wage rates than the employed. More importantly, this informal income is seldom sufficient on its own to enable unemployed households to get by, although it is a useful supplement to make ends meet. This is shown in Italy (Mingione 1991), the UK (Bryson and Jacobs 1992, Evason and Woods 1995, Howe 1988, Jordan et al. 1992, MacDonald 1994, Morris 1993, Pahl 1984), the Netherlands (Renooy 1990, Van Eck and Kazemeier 1985), the US (Jensen et al. 1995) and Canada (Fortin et al. 1996, Lemieux et al. 1994). Indeed, many unemployed people can only accept such low-paid informal jobs because they have the safety cushion of their benefits (Howe 1990, Jessen et al. 1987, Leonard 1994). This is a crucial finding for it shows that informal employment is not an alternative to state assistance as some have suggested (see chapter 10).

In sum, these studies reveal that the employed constitute a greater proportion of the informal labour force than the unemployed but that certain populations of the unemployed make more use of informal employment than others. Moreover, the employed tend to undertake more autonomous, creative and personally rewarding informal employment whilst the unemployed tend to engage in low-paid, repetitive, monotonous and exploitative work. The result is that informal employment usually reinforces, rather than mitigates, the plight of the unemployed. A segmented informal labour market can be identified in which there is a well-paid 'core' informal workforce of people who often also have a job and a 'peripheral' informal workforce of frequently unemployed or underemployed people who tend to engage in the more exploitative lower-paid forms of such work when they are successful at gaining entry into this informal labour market.

Explaining the extent and nature of participation in informal employment amongst the employed and unemployed

From our analysis of the one-off locality studies of informal employment, we have deduced that explanations for the nature and extent of the participation of the unemployed and employed in informal employment have to be sought through an analysis of the ways in which a range of economic, social, institutional and environmental conditions combine in a locality to produce a particular configuration of informal employment. Hence, it is not the existence of a specific condition on its own which causes a particular configuration of informal employment amongst the unemployed in a locality but, rather, it is the way in which various conditions combine to produce an area conducive or otherwise to the participation of the unemployed in informal employment.

To explain the fact that many unemployed fail to participate in informal employment and when they do, they engage in 'peripheral' informal work, five conditions are commonly cited. These constitute the 'cocktail' of factors that combine in a wide range of localities in the advanced economies to produce a barrier to entry to informal employment for the unemployed. First, there is the economic factor that the unemployed lack the money to acquire the goods and resources necessary to engage in informal employment (Miles 1983, Pahl 1984, Smith 1986, Thomas 1992). For example, without access to a car, they cannot always travel to where the work is available, whilst without the tools necessary to undertake the work (e.g. ladders, workbenches, workrooms and equipment), they cannot conduct a wide range of paid informal activities. The result is that the unemployed have fewer opportunities than the employed to engage in autonomous informal employment or even gain access to organised work.

Second, there is the social factor that the reduction in the size of social networks following redundancy means that the unemployed have fewer opportunities for undertaking work or for receiving help (Engbersen *et al.* 1993, Howe 1990, Miles 1983, Mingione 1991, Morris 1993, 1995, Renooy 1990, Thomas 1992). Given that the long-term unemployed, moreover, mix mostly with other long-term unemployed, have relatively few friends or acquaintances who are employed (Morris 1995) and that the majority of informal employment is found through acquaintances and employment (Van Eck and Kazemeier 1985), the result is that the unemployed again have fewer opportunities for hearing about informal employment than those in employment, especially opportunities for autonomous work. As Komter (1996) finds in the Netherlands, and contrary to popular belief, not only are the social networks of the employed and higher-educated often more numerous and extensive than those of the unemployed and lesser-educated, but informal exchange is used as a way of maintaining them. Similar findings have been identified on new estates in France where informal exchange is undertaken as much for social as for economic reasons (Cornuel and Duriez 1985).

A third reason given for the relatively low participation of the unemployed in informal employment, which is again a social factor, is asserted to be their lack of skills (Fortin *et al.* 1996, Howe 1990, Lysestol 1995, Mingione 1991, Renooy 1990, Smith 1986). If their skills are inappropriate for finding formal employment, there is no reason to believe that they can sell them on the informal labour market either in an autonomous or organised form. So, besides lacking economic and social capital, it is argued that many of the unemployed also lack the human capital necessary to engage in informal employment. The consequence is that they are relegated to engaging only in unskilled or semi-skilled informal employment. Having a formal job, moreover, means that the outside world recognises a person as having a skill to offer and is a legitimisation of these skills in the eyes of potential customers for informal work.

A fourth reason for the lack of participation of the unemployed, particularly claimants, in informal employment is asserted to be that they feel more inhibited about engaging in such work for fear of being reported to the authorities and having their benefit curtailed. Given that working informally whilst claiming social security is considered in the advanced economies to be a more serious offence than engaging in tax fraud (Aitken and Bonneville 1980, Cook 1997, Deane and Melrose 1996, Jordan *et al.* 1992, Keenan and Dean 1980, Weatherley 1993), the result is that the claimant unemployed are less likely to engage in informal employment than the employed. This 'institutional' factor is felt to be particularly relevant in welfare regimes which have both universal and comprehensive social security benefits as well as strict enforcement of social security regulations, such as Germany, Norway and Denmark (Hellberger and Schwarze 1987, Lysestol 1995, Mogensen 1990, Wenig 1990).

Finally, there is the environmental or geographical factor that a greater proportion of the unemployed live in areas disrupted by economic restructuring and crisis (Morris 1994, Pahl 1984, Smith 1986). Such areas are said to fail to provide not only formal but also informal employment. There is little point, for example, being a window-cleaner in such areas if nobody can afford to have their windows cleaned (Coffield *et al.* 1983). As such, and given the spatial concentration of the unemployed (e.g. Dorling and Woodward 1996), there is simply less demand for informal employment in these areas than in others.

Therefore, the unemployed possess the free time but lack the additional resources and opportunities necessary to conduct a wide array of informal employment. Conversely, the employed have fewer fears about the authorities, have the money to acquire the necessary tools and materials, wider social networks, more chance of hearing about informal employment and greater scope for undertaking such work in the areas in which they live. Pahl (1988: 255) refers to this as 'the Matthew effect': to them that have, more is given, whereas to them that have little, even that which they have is taken away.

Such findings and explanations have widespread appeal for many commentators who discuss the configuration of the informal labour market. These studies show that such work cannot be construed as simply a substitute to which the unemployed turn when excluded from the formal labour market and neither can it be used as an excuse for cutting back on social security contributions to the unemployed. The problem, however, is that although the vast majority of studies conform to this finding about the relationship between unemployment and informal work, the evidence is not unequivocal. Put another way, although the vast majority of localities studied possess this 'cocktail' of factors which prevent the participation of the unemployed in informal employment, there are other studies of different localities which seem to suggest that alternative 'cocktails' can exist which offer the unemployed the opportunity to engage as much if not more than the employed in such work.

Alternative studies, that is, suggest that similar levels of informal employment are to be found amongst the employed and unemployed (Ferman *et al.* 1978, Hellberger and Schwarze 1987, Mogensen 1990, Wenig 1990). In Germany, for example, Wenig (1990) and Hellberger and Schwarze (1987) find that the unemployed are just as likely to engage in informal employment as the employed in that 9.3 per cent of each group participate in such work. Yet other locality studies, moreover, reveal that the unemployed engage in more informal employment than the employed. This is identified to be the case in Belfast (Howe 1988, Leonard 1994) and Belgium (Kesteloot and Meert 1994, Pestieau 1984). Pestieau (1984), for example, in the Liege area of Belgium, identifies that suppliers of informal labour were especially likely to be found among manual workers, unemployed or unoccupied persons, low-income and uneducated groups and younger people. A similar configuration is also identified in Spain (Miguelez and Recio 1986, Lobo 1990a) and Greece (Hadjimichalis and Vaiou 1989, Leontidou 1993), albeit only so far as organised informal employment is concerned.

Not only is there empirical data, therefore, to support alternative configurations of informal employment, but there is also evidence that some of the conditions used by analysts to explain the low level of participation amongst the unemployed are not universally applicable. Take, for example, the issue that benefit fraud is everywhere seen as more unacceptable than tax fraud. The World Values Survey conducted between 1981 and 1983 clearly shows that this is not the case (see Table 4.2). In the US, Britain and Australia, tax fraud was seen to be either as equally unjustifiable as benefit fraud or more justifiable. However, in Sweden, Germany, Canada and France, the reverse was true – tax fraud was seen as less justifiable than benefit fraud, reflecting differences in benefit system structures, culture, beliefs and politics.

Table 4.2 Percentage of public saying that benefit and tax fraud is never justified

	Benefit fraud	Tax fraud
Sweden	82	90
USA	76	75
Britain	75	72
Australia	74	49
Germany	63	76
Canada	61	73
France	40	46

Source: World Values Survey 1981–83, cited in Weatherley (1993).

There is a need, therefore, to move beyond the previously over-simplistic generalisations and explanations concerning the participation of the unemployed in informal employment towards more context-bound understandings and explanations.

If we examine closely the explanations cited above of why the unemployed do not engage in informal employment, it is clear that they do not necessarily apply to all unemployed everywhere (e.g. few skills, no social networks, living in poor areas, afraid of the authorities). These are conditions frequently related to unemployment, but not always part and parcel of the experiences of all of the unemployed in every locality. Here, we attempt to explain the exceptions to the rule about the participation of the unemployed in informal employment by taking a case study of a working-class housing estate in Catholic West Belfast studied by Leonard (1994). The aim in so doing is to further explain why the relationship between informal employment and employment status is place-specific rather than universal.

A case study of West Belfast

In this area of West Belfast, characterised by widespread long-term unemployment, Leonard (1994) discovers that in contrast to many of the findings cited above, informal employment is rife and that much of it is undertaken by the unemployed. As Table 4.3 shows, of the ninety-three married couples questioned, none of the dual-earners engaged in informal employment, but of the couples in which both were unemployed, 49 per cent participated in 'organised' and 58 per cent in 'individual' informal employment. Amongst the fifty-seven single, divorced and widowed respondents, similarly, none of the employed but 15 per cent of the unemployed engaged in organised informal employment, whilst 33 per cent of the employed compared with 8 per cent of the unemployed engaged in individual informal employment. On the whole, therefore, the unemployed participate in informal employment on this estate more than the employed and the work which they undertake is of both the individual and organised kind.

How, therefore, can this exception to the rule be explained? What are the conditions that have caused such a configuration? Here, we review how the various conditions which shape the extent and nature of informal employment discussed in chapter 3 have combined in a particular way in this area to cause this specific configuration of the local informal labour market.

Starting with the economic conditions, the high level and long duration of unemployment has been an important factor on this West Belfast estate in shaping the structure of the informal labour market. This is because unemployment is a product of endemic economic crisis, rather than merely the temporary result of cyclical downturns. This is not only due to global economic restructuring, but also to what have been until recently seen as the seemingly intractable political problems which beset Northern Ireland. With little likelihood of

Table 4.3 Distribution of informal employment in West Belfast: by household composition and employment status

Household composition	% engaging in:	
	Organised informal employment	Autonomous informal employment
Married couples (N = 93):		
Husband and wife formally employed (N = 10)	0	0
Husband or wife formally employed (N = 26)	0	27
Husband unemployed and wife housewife (N = 45)	49	58
Husband and wife retired (N = 12)	8	17
Single, divorced and widowed respondents (N = 57):		
Formally employed (N = 6)	0	33
Housewife (N = 22)	4	18
Unemployed (N = 13)	15	8
Retired (N = 16)	0	0

Source: Derived from Leonard (1994: Table 9.1).

finding formal employment in the near future or beyond, the unemployed have thus sought informal employment, reinforcing Gallie's (1985: 522) assertion that 'the most probable location of the growth of widespread informal work will be in areas of catastrophic economic collapse where the possibility of re-insertion into normal economic activity appears particularly remote'. However, this condition, although necessary, is insufficient alone to cause the participation of the unemployed in informal employment in the area. Indeed, in many areas with high levels of very long-term unemployment, such as Hartlepool (Morris 1994), the unemployed tend not to conduct such work. Moreover, in some areas with lower unemployment levels and less long-term unemployment, such as Amsterdam (Renooy 1990), the participation of the unemployed in informal employment appears to be as extensive as, or indeed more than, that in deprived areas. So, the level and duration of unemployment has different impacts in different areas due to the host of additional social, environmental and institutional characteristics which prevail.

Indeed, the social conditions on this Belfast estate are an important determinant of the participation of the unemployed in such work. First, there is the nature of the social networks. In areas where social networks are created mainly through the employment-place, unemployment causes a contraction of social contacts and thus of opportunities for hearing about informal employment (Barthe 1988, Foudi *et al.* 1982, Morris 1994). Here, in West Belfast, however, there are strong alternative social networks based outside the employment-place in the form of extensive kinship and friendship networks. This is a product of the social and spatial isolation of the estate, the lack of in- and out-migration and the social homogeneity of the population. The result is that the population is not dependent upon the employment-place for hearing about opportunities for informal employment. Hence, dense social networks here mitigate the deprivation and high unemployment that might elsewhere reduce opportunities for informal employment. Social networks, however, are only communication channels that can put the unemployed in touch with work opportunities. If no such opportunities exist, then dense social networks are of no value: they have nothing to communicate.

A second social condition influencing the participation of the unemployed is the socio-economic mix of the area. Most studies reveal that it is when an area unites people with high incomes but little free time with others who have low incomes but much free time, that the unemployed engage in greater amounts of informal employment (Barthelemy 1990, De Klerk and Vijgen 1985, Pestieau 1984, Renooy 1990, Sassen 1991, Terhorst and Van de Ven 1985). This West Belfast estate, however, is a socially homogeneous area which, in this instance, aids and abets informal employment because the common deprivation shared by most residents makes the undertaking of such work an acceptable activity, as do the 'political' values of the estate, discussed further below. Due to widespread deprivation, however, prices of, and incomes from informal employment are fixed at a level somewhat lower than the formal sphere so as to allow exchange to take place. Although the income from informal employment may not allow many purchases to be made outside this sphere, it does provide the means to 'buy back' in a reciprocal manner other informal goods and services.

A final social characteristic which is of major importance in shaping the extent of informal employment on this estate is the local cultural traditions and the consequent social norms and moralities. The unemployed in this locality demonstrate little fear about being caught 'doing the double', that is, undertaking informal employment whilst claiming social benefits. This is not the case elsewhere in the northern EU nations (see, for example, Wenig 1990). Here, there is less fear of detection because the local social mores are not conducive to reporting such behaviour. In Catholic West Belfast, as Howe (1988) has highlighted, there is a greater moral acceptance of working whilst claiming benefits than in Protestant East Belfast, in part because the legitimacy of the state is less recognised in the former area. When the state is not even recognised as a valid institution amongst a proportion of the population and a 'shadow' state is in operation, it is little surprise to find people, whether employed or not, engaging in informal employment.

It is not only the republican issue, however, which makes the participation of the unemployed in informal employment socially acceptable on this estate. As other studies show, such conducive social mores also result from the widespread deprivation, social isolation and/or the social cohesiveness of the population (Barthelemy 1991, Legrain 1982, Noble and Turner 1985, Weber 1989). Here, and akin to other similar areas, informal employment is perceived as an acceptable part of community life rather than a tax or social security fraud and is undertaken as much for social as for economic reasons. Consequently, the monetary aspect of the activity is suppressed in the minds of those engaged in it and thus so is its illegality.

Besides the social characteristics of this Belfast estate influencing the informal employment of the unemployed, there are also institutional factors. It is not only the way in which rules and regulations are interpreted and enforced by the population itself, but also by the state, which sway the extent of the informal employment conducted by the unemployed. In some areas of southern EU nations, for example, lax enforcement of state rules and regulations is often deliberately promoted to help the unemployed engage in informal employment (Portes and Sassen-Koob 1987, Warren 1994). In northern EU nations, however, measures to catch any unemployed engaging in informal employment are strictly enforced (Pahl 1990, Wenig 1990). The problem is that there are particular difficulties in implementing them on this estate and more subtle forms of investigation have had to be introduced (Leonard 1994). The social cohesion of the estate, none the less, allows the usual institutional constraints that prevent the unemployed from engaging in informal employment to be negated.

Finally, the nature of the physical environment can influence the participation of the unemployed in such work. It is the relative isolation of this Belfast estate and the poor access to formal services which has led to many forms of informal employment developing, such as 'house shops' and 'community bakers'. Such work provides access to goods and services that would not otherwise be available, thereby ameliorating deprivation, even if not providing a complete survival strategy for the unemployed. Legrain (1982) finds the same in France in Grand Failly, a steel town set in the context of a traditional rural community, as does Renooy (1990) amongst the coalfield communities in the Netherlands.

Conclusions

This chapter has shown that the unemployed do not disproportionately participate in and gain from informal employment and that such activity is not therefore primarily a survival strategy used universally or even in large part by the economically excluded. Instead, the unemployed not only undertake less informal employment than the employed but the work which they do undertake is more exploitative, lower-paid and less fulfilling than that conducted by the employed. In consequence, a segmented informal labour market can be conceptualised in which there is a well-paid core informal workforce of people who frequently also have a formal job and a peripheral informal workforce of often unemployed or underemployed people who tend to conduct the more exploitative lower-paid forms of such work when they are successful at gaining entry into this informal labour market at all.

However, as we have seen above, this state of affairs does not have a single cause. It results from the combination of a number of factors that come together in many, although not all, areas and affect large proportions of unemployed people but not the unemployed universally. In other words, within certain populations, the economic, social, institutional and environmental conditions which structure the lives of the unemployed diverge from the 'norm' which usually excludes them from informal employment. Nevertheless, it must be stressed that the dominant way in which the economic, social, institutional and environmental characteristics of populations combine in most localities serves merely to reinforce, rather than ameliorate, the plight of the unemployed. Given that informal employment is currently configured in such a manner, a key question to address is whether such employment can be reconstructed in a manner which reverses such disparities and provides the unemployed with an alternative coping strategy. It is to this issue which we shall return in Part III of this book.

5

GENDER AND INFORMAL EMPLOYMENT

Introduction

The aim of this chapter is to fill a major gap in the literature on gender divisions of labour. Although many studies have examined the gender segmentation of employment and/or gender divisions of unpaid work (e.g. Gershuny *et al.* 1994, Gregory and Windebank forthcoming, Hakim 1995, Martin and Roberts 1984, Vogler 1994, Walby 1997, Warde and Hetherington 1993), few if any have explored the allocation of informal employment by gender.[1] To examine this missing work in the study of gender divisions of labour, we now analyse the extent and character of women's and men's participation in informal employment, revealing the nature of the gender segmentation of such employment. Subsequently, we examine the various economic, social and institutional characteristics that shape the gendered nature of informal employment.

Extent of participation in informal employment by gender

Are women more likely to participate in informal employment than men or vice versa? For some, the assumption is that such employment provides comparatively greater opportunities for women than men (e.g. Priest 1994). However, as with our findings in the last chapter concerning the unemployed, the overarching finding of the vast majority of studies conducted on informal employment is that the extent of women's participation is less than that of men and that men constitute the majority of the informal labour force. This is identified to be the case in the Netherlands (Van Eck and Kazemeier 1989, Renooy 1990), the UK (MacDonald 1994, Pahl 1984), Italy (Mingione 1991, Vinay 1987), Denmark (Mogensen 1985), the United States (McInnis-Dittrich 1995) and Canada (Lemieux *et al.* 1994, Fortin *et al.* 1996). It is important to state, nevertheless, that the relative gap in the participation rates of men and women is not great. For example, in three Canadian regions, Fortin *et al.* (1996) find that men constitute just over half of all informal workers

Table 5.1 Informal wage rates in three regions of Canada: by gender, 1993

	Quebec (1985 in parentheses)	Montreal	Bas-du-Fleuve
% engaging in informal employment:			
Men	5.4 (6.4)	6.6	4.1
Women	4.2 (5.6)	5.5	3.2
% of all informal workers:			
Men	52.6 (51.8)	52.1	51.3
Women	47.4 (48.2)	47.9	48.7
Average hours p.a. (C$):			
Men	403 (275)	581	309
Women	417 (450)	614	317
Average salary p.a. (C$):			
Men	3,093 (2,744)	5,136	2,850
Women	3,994 (2,444)	3,949	1,652
Average wage/hour:			
Men	7.67 (9.98)	8.83	9.22
Women	9.57 (5.43)	6.43	5.21

Source: Derived from Fortin *et al.* (1996: Tables 3.1–3.3).

(51.3–52.6 per cent) and about 1 per cent more men than women engage in informal employment (see Table 5.1). Similar sized gaps in participation rates are also identified in Denmark (Mogensen 1985), the Netherlands (Van Eck and Kazemeier 1989) and Italy (Mingione 1991). This difference in men's and women's participation rates in informal employment is perhaps not so surprising as it first appears when it is realised that men more frequently engage in paid employment than women in nearly every context. Indeed, if anything, the gap in informal compared with formal participation rates is narrower, revealing that women do indeed find it easier to work informally than formally.

Exceptions to this general rule, nevertheless, do exist. Lobo (1990a and b), for example, argues that women rather than men are more likely to be engaged in informal employment in Portugal and Spain. This may be correct but it is probably because of the cocktail of factors present in these areas and the concomitant fact that these studies only consider one limited type of informal employment: the low-paid, organised, exploitative, informal employment which, as we shall now show, is where women's informal jobs are concentrated. Furthermore, and as we shall see later, in some geographical contexts, the flexibility which is characteristic of women's employment is expressed in terms of informal employment rather than formal part-time or short-term contract employment.

Character of informal employment by gender

To examine the character of the informal employment undertaken by men and women, we examine first, the types of informal employment undertaken, second, the different motivations of men and women for engaging in such work and third, the differentials in their pay rates.

Types of informal employment undertaken by men and women

Examining the types of informal employment men and women perform, it is striking that women undertake a very different set of tasks than men. Just as there is a clear gender segmentation by sector in the formal labour market with women heavily concentrated in service sector jobs (e.g. Townsend 1997), the same is true of the informal labour market. As Hellberger and Schwarze (1986) identify in Germany, whilst 12.3 per cent of informal workers are in the primary sector, 35.8 per cent in manufacturing and 51.9 per cent in services, these figures are 6.2 per cent, 11.8 per cent and 82.0 per cent for women (and 15.7 per cent, 49.3 per cent and 35.0 per cent for men). It is similar elsewhere. Women tend to engage in service activities such as commercial cleaning, domestic help, child-care and cooking when they undertake informal employment. Men, on the other hand, tend to engage in what are conventionally seen as 'masculine tasks' such as building and repair work (Fortin *et al.* 1996, Jensen *et al.* 1995, Leonard 1994, Mingione 1991, Pahl 1984). In other words, women conduct activities in informal employment which mirror their domestic role, perhaps even more so than the distribution of women's employment in the formal sphere. The informal labour market, therefore, reflects and re-inforces the gender divisions in activities prevalent in both the formal labour market and unpaid work. Related to this, women are often seen to engage in what are traditionally defined as low-skilled, exploitative and organised forms of informal employment whilst men conduct work viewed as higher-skilled, more rewarding and autonomous (Leonard 1994, McInnis-Dittrich 1995, Phizacklea and Wolkowitz 1995, Prugl 1996).[2]

It is not only the types of informal activities that differentiate men from women. There is also tentative evidence that in some particular contexts men's participation in informal employment is more infrequent but full-time than women's, which although continuous tends to be part-time (Leonard 1994, McInnis-Dittrich 1995). As Leonard (1994: 162) finds in Belfast, 'While the women were usually in constant informal employment compared to the men, whose employment was more casual, nonetheless . . . the women tended to work part-time while the men tended to work full-time'. In Appalachia, meanwhile, McInnis-Dittrich (1995) arrives at the same conclusion. Women engaged in regular small part-time informal jobs such as housework, consign-ment quilting, gardening, child- or elder-care, yard sales, aluminium recycling or work in the tobacco fields, whilst men tended to engage in more irregular full-time informal employment, such as in the local sawmill, the sanitary landfill or on the farm. So, not only is informal employment sectorally divided in similar ways to formal employment but the part-time/full-time dichotomy which is often prevalent in formal employment also appears to be replicated in the informal labour market.

There are also major differences in the formal employment status of men and women who conduct such work. So far as men are concerned, and as shown in the last chapter, most evidence suggests that informal employees tend to be formally employed rather than unemployed. For women, however, it is the non-employed who do much of this work. Take, for example, the study by Pahl (1984) on the Isle of Sheppey. When the twenty-seven informal workers identified in the study were analysed, some salient gender contrasts emerged. Of the eleven men, ten were in full-time employment. Only one was un-employed. By contrast, of the sixteen women, eight were full-time housewives. This difference is largely related to the type of work undertaken. Men, in general, are more likely to do home improvements, which if unemployed is conspicuous to observers, whilst women overwhelmingly provide routine domestic services, which are not so noticeable and are often perceived rightly or wrongly as unpaid activities. In addition, 'full-time housewives' are one step removed from the authorities in the sense that they do not claim benefit them-selves, sign on, get called for interviews or even register as unemployed. This unequal evaluation of various tasks is to men's advantage when they are in formal employment but is held against them if they become unemployed and seek to work informally (Pahl 1984). Furthermore, most informal employment involving men requires skills which unemployed men are viewed as unlikely to possess whilst most informal employment for women uses skills which most women are perceived to possess (Renooy 1990), since these emanate from the family and household, not the employment-place. Indeed, these women are often acting as substitute 'housewives' for other women with more lucrative employment.

In sum, the informal labour market is segmented along gender lines both sectorally as well as by employment pattern and the employment status of men and women who engage in such work. As we shall now see, this is because women have to engage in informal employment which reflects their domestic roles (resulting in sectoral divisions) or which fits in with their domestic duties (resulting in the tendency for regular but part-time informal employment so that family needs can be met).

Motivations for conducting informal employment: gender differences

As shown above, women undertake informal employment mainly when the household needs to generate extra cash and when it is the only work available which fits in with their domestic caring responsibilities, such as for children or elderly relatives. Men, on the other hand, undertake informal employment for very different reasons. For them, it is much more about generating spare cash or pocket money to finance social activities and to differentiate themselves from the domestic realm and women (Leonard 1994, MacDonald 1994, Morris 1987, 1995).

Take, for example, the studies by Morris (1987, 1995) in South Wales and Hartlepool. For men, informal employment is found to fulfil two social functions: it provides a means of disassociation and differentiation from the domestic sphere and women; and acts as a source of additional earnings which finance a degree of social activity, notably drinking. Informal employment for women, meanwhile, in so far as it exists, relates to the conventional gender division of labour and to established norms about gender roles and identities. 'Thus female spending patterns are bound with their association with the domestic sphere, and to be contrasted with men's perceived need for a "public" identity achieved through social spending' (Morris 1987: 100–1). Leonard (1994) finds similar differences in the uses to which informal earnings are put by men and women in Belfast. Women tend to regard their income as the family's wage and use their earnings to provide for the family's everyday needs. Men, on the other hand, use either part or all of the income derived from working informally to satisfy their personal needs. Some men felt that to hand over the whole informal income to satisfy family needs would mean that they were working for nothing. In this sense, women use such income for primarily economic reasons whilst men use it more for social reasons.

However, such a gendered division of the motivations for conducting informal employment is not so exclusive as suggested above. As Leonard (1994) highlights, women also use such work as a tool to reconstitute continually their social networks. For example, catalogue buying and selling gave women an opportunity to forge networks and cement personal and social ties. Transactions were highly socialised, giving women the opportunity to network with other housewives on the estate. However, and as Leonard (1994) asserts, these personal social relationships between women on the estate were often also manipulated for entrepreneurial advantage. Building up friendships and relationships with others on the estate was frequently a cultural device, strategically used to make money. Similarly, for men, although such work might be undertaken to fund socialising, this can also be seen as a mixture of economic and social reasons since social interaction is seen by many of these men as a basic need and a source of further work.

Gender variations in informal wage rates

The overwhelming finding of the vast majority of studies of informal employment is that women generally earn lower informal wages than men (Fortin *et al.*

70

1996, Hellberger and Schwarze 1986, Lemieux *et al.* 1984, McInnis-Dittrich 1995). Take, for example, the studies by Fortin *et al.* (1996). As Table 5.1 shows, men who engage in informal employment work fewer hours than women and have higher average total incomes from such work in both Montreal and Bas-du-Fleuve. The result is that the average hourly wage of women informal workers is lower than that of men. Although this was also the case in Quebec in 1985, it was not the situation by 1993. Despite men continuing to be overrepresented in the informal labour force in 1993, the average number of hours worked by male informal labourers increased considerably between 1985 and 1993 compared with a decline in the average hours of women, yet the average hourly male informal wage drastically reduced whilst for women it underwent a considerable increase. Indeed, by 1993, the average hourly wage of women was higher than that of men. Although Fortin *et al.* (1996) neither identify nor explain this shift in their original analysis, personal correspondence with the authors reveals that they consider this to be due to the recession in Quebec in the early 1990s, which hit men's activities particularly, such as work in construction and repair, whilst the ongoing feminisation of the formal labour force is likely to have led to a shortage of informal labour in what are traditionally perceived as feminine tasks, such as child-care and domestic cleaning, leading to a rise in informal wage rates for conventional women's work. This important exception, nevertheless, despite showing that there are particular economic and social circumstances in which women can and do receive higher informal wage rates than men, should not distract attention from the fact that women generally earn less than men in the informal labour market.

Moreover, and as Table 5.2 reveals in Germany, although women dominate the lower-paid echelons and men the higher-paid spheres of informal employment, there are men in particular places and circumstances who tend to earn a very poor wage from their informal employment and women who earn a relatively high hourly informal wage. For example, in areas where child-care is scarce, one would expect relatively high informal incomes from it. Indeed, rates of pay for childminding vary enormously from one suburb to another, let alone one city to another. Indeed, this caution in allocating poorly and well-paid informal employment exclusively to women and men respectively is reinforced by findings in Italy. Here, research indicates that some women do fare relatively well from their informal employment, especially in the Red regions (Cappechi 1989, Vinay 1985), such as those women in Umbria who fabricate mini-motors for tele-printers in their basements (Mattera 1985). Consequently, it would be erroneous to characterise the segmented nature of the informal labour market in too rigidly gendered terms. Nevertheless, as in the formal labour market, the existence of women in executive positions does not undermine the general trend of women being ghettoised at the lower end of the employment hierarchy.

Table 5.2 Average hourly wage rates in second jobs, Germany

DM per hour	% of second jobs in this income bracket	% of men in second jobs	% of women in second jobs
<5	7.4	7.1	7.9
5–8	11.7	3.9	28.5
8–10	10.2	9.2	11.9
10–12	21.9	24.9	14.1
12–15	10.7	12.9	6.6
>15	38.1	42.0	31.0

Source: Hellberger and Schwarze (1986: Table 2).

Explanation for the gender configuration of informal employment

To explain the gender configuration of informal employment, it is necessary to understand the broader economic, social and institutional context within which such activity takes place in the contemporary advanced economies. To do this, we examine the ways in which the economic characteristics of places influence the gendered configuration of informal employment. We then examine the range of social characteristics that combine with these economic characteristics to produce specific configurations and finally, we examine the ways in which institutional characteristics can mediate these economic and social characteristics in specific ways in different circumstances to produce particular configurations. The result is an explanation of the various configurations of the gendered nature of informal employment to be found in different places.

Economic characteristics

One of the principal socio-economic changes having taken place in the advanced economies over the past three decades has been the feminisation of the labour force. Whilst the proportion of working-age women in formal jobs has steadily grown, the share of prime-age males in employment has rapidly declined (e.g. International Labour Organisation 1996, Townsend 1997, Walby 1997). If one scratches the surface of this seemingly nascent labour market equality, however, it rapidly becomes apparent that traditional gender disparities remain. Numerous studies reveal that women in the advanced economies are more likely than men to be employed in atypical forms of employment (e.g. part-time and temporary work), to find themselves at the bottom of the authority and skills hierarchy and to be low-paid (Briar 1997, Townsend 1997, Walby 1997). The result is that more women live in poverty than men (Baxter 1992, Beneria and Stimpson 1987, Gornick and Jacobs 1996, Hanson and Pratt 1995, Kiernan 1992). It is unsurprising, therefore, that when women do engage in informal employment, it is of a kind which can be described as an extension of the atypical, flexible jobs to be found in the formal labour market.

Over the past fifteen years, a plethora of studies has highlighted the national and regional variations which exist in levels and patterns of women's formal employment (Eurostat 1996, Gregory and Windebank forthcoming, Massey 1984, Walby 1997). Little cross-national research has been conducted, however, on women's participation in informal employment. Nevertheless, given our nationally-specific understanding of women's paid informal activities, it is perhaps safe to assume that in countries and regions in which women have access to formal employment that they can fit with their familial responsibilities (that is, either on a full-time basis accompanied by adequate child-care, or on a part-time basis which compensates for insufficient or costly external child-care possibilities), they will not have recourse to informal employment. Indeed, if formally employed, it is hard to imagine that women with caring responsibilities could or would wish to accumulate informal employment on top of their formal jobs and unpaid domestic work (see discussion of 'social characteristics' below). However, where access to such formal employment is limited, either because of the generally poor state of the local labour market or because those formal jobs available do not give scope for women to combine them with their caring responsibilities, they are more likely to engage in informal employment. Therefore, and as we shall see below, social factors and institutional factors have an important part to play alongside economic conditions in structuring women's participation in informal employment.

Social characteristics

Economic restructuring shapes, and is shaped by, the social structure of an area and it is how these two processes combine in a particular place which is an important determinant of the gender configuration of informal employment. Here, we are referring principally to local cultural traditions, social norms and moralities concerning women's participation in the formal labour market and men's participation in the domestic division of labour. As shown above, the extent and nature of women's participation in informal employment either mirrors their domestic roles or fits in with their domestic duties and responsibilities, if not both. It is important, therefore, in any consideration of women's employment outside the home, whether formal or informal, to examine the nature of their domestic workload and the extent to which this is shared with men. It might be suggested that if these domestic roles and duties are changing or are different in different places, then the configuration of informal employment will vary both temporally and spatially.[3] To what extent, therefore, do women remain responsible for the domestic workload? Is there any evidence that this is different in different places? Is this division of domestic labour changing over time?

On the issue of whether women remain responsible for the domestic work-load and whether this is different in different places, a cross-national series of time-budget studies reveals that despite rhetoric about the 'new man', there has been little renegotiation of the gender division of domestic labour across the advanced economies. Table 5.3 shows the extent to which engaging in employment reduces the participation of women in domestic work. When women are employed part-time, there is only a slight reduction in time spent on domestic work as compared with non-employed women across all of the

Table 5.3 Structure of by gender and employment status: by country (in hours per week)

	UK 1984	Canada 1981	Denmark 1975	France 1986	Norway 1981	Holland 1980	USA 1985
MEN							
Full-time employed:							
Personal	73.5	70.4	70.7	71.0	70.9	70.4	71.0
Professional	43.4	40.5	44.7	44.9	42.6	41.6	47.5
Domestic	13.0	16.8	6.4	16.5	14.7	9.7	14.1
Free	38.1	40.3	46.2	35.6	39.8	46.3	35.4
Part-time employed:							
Personal	74.0	71.1	71.3	74.0	75.3	78.8	78.1
Professional	27.6	25.9	33.5	33.3	24.9	8.7	16.2
Domestic	18.1	15.0	8.1	13.4	12.4	9.7	18.2
Free	48.3	56.0	55.1	47.3	55.4	70.8	55.5
Non-employed:							
Personal	81.3	75.4	80.5	80.2	80.9	80.4	81.0
Professional	2.6	3.4	9.0	1.4	1.7	0.4	6.2
Domestic	20.9	19.5	11.1	23.6	17.1	20.8	23.2
Free	63.2	69.7	67.4	62.8	68.3	66.4	57.6
WOMEN							
Full-time employed:							
Personal	73.9	72.7	70.0	72.5	69.7	73.6	72.3
Professional	40.1	35.6	39.0	40.1	36.0	31.4	36.3
Domestic	18.9	22.9	15.8	27.7	24.7	13.6	25.8
Free	35.1	36.8	43.2	27.7	37.6	49.4	33.6
Part-time employed:							
Personal	75.0	70.7	74.3	74.1	73.1	76.8	77.1
Professional	20.1	24.5	21.6	25.8	18.7	11.1	14.8
Domestic	36.9	32.8	26.6	36.0	35.1	23.5	35.7
Free	36.0	40.0	45.5	32.1	41.1	56.6	40.4
Non-employed:							
Personal	79.6	75.1	79.8	79.7	76.5	77.8	78.1
Professional	1.6	1.8	3.3	0.8	1.7	0.7	2.8
Domestic	37.5	35.2	28.1	40.0	38.2	41.5	38.2
Free	49.3	55.9	56.8	47.5	51.6	48.0	48.9

Source: Roy (1991: Table 1).

nations studied. This is because part-time employment is undertaken by women precisely so that such work can fit in around their domestic roles and duties (e.g. Beechey and Perkins 1987, Walby 1997). For full-time employed women, meanwhile, although the reduction in their domestic workload is greater than for part-time employed women, this is again insufficient to compensate for the extra hours that they spend in employment. As women enter employment, therefore, their total workload (domestic plus employment time) increases, resulting in a 'double burden' of domestic work and employment for these women (e.g. Gershuny et al. 1994).

Indeed, even when a man becomes unemployed, studies have found little renegotiation of the domestic division of labour and sometimes a decrease in men's contribution (Bell and McKee 1985, Pahl 1984, Pinch and Storey 1992). In part, this is due to women not wanting to further harm their man's self-esteem by asking for a contribution to housework, which indirectly displays how devalued domestic work is in contemporary society. Similarly, when couples retire, little renegotiation of the gender division of household tasks occurs for much the same reasons (e.g. Szinivacz and Harpster 1994).

So, men conduct a very small proportion of the domestic work and their total workload is universally much lower when men and women are compared by their employment status. Nevertheless, such statistics do not tell the whole story. They obscure the fact that much of men's domestic work is non-routine, such as gardening and DIY projects. Indeed, it could be suggested that some of the autonomous informal employment undertaken by men derives from the skills which they acquire in such masculine forms of self-provisioning. Men participate less, however, in routine housework tasks (e.g. cleaning, cooking, ironing) and more frequently 'help out' with, rather than 'take responsibility' for, the domestic workload (Brannen et al. 1994, Brannen and Moss 1991, Gershuny et al. 1986, 1994, Kiernan 1992, Pinch and Storey 1992).

Such findings have important inferences for the gender configuration of informal employment. Women remain everywhere primarily responsible for unpaid domestic work, despite their increasing insertion into the formal labour market. The result is that they have little extra time for engaging in informal employment. Formally employed women with familial responsibilities in particular will thus rarely engage in informal employment. In contrast, given that the traditional 'male the breadwinner' and 'woman the homemaker' model is not dead, it is not surprising that men take on informal employment to earn additional family income. When women do enter the informal labour market, moreover, the lack of any fundamental renegotiation of women's roles and duties is mirrored in the character of their activity. They undertake tasks traditionally associated with the domestic domain, frequently on a part-time basis so as to enable them to fulfil their domestic roles, and the principal motivation for their participation is to meet household needs.

Is there any evidence that this domestic division of labour is changing over time and thus that the gender division of informal employment may be

Table 5.4 Change in work time in married couples in eight countries, 1961–90

	Men's time as a % of total for: All work-related activities (excluding commuting)				All unpaid work			
	1961–70	1971–77	1978–82	1983–90	1961–70	1971–77	1978–82	1983–90
Canada	–	–	57	–	–	34	37	–
France	54	52	–	–	25	28	–	41
Netherlands	–	59	57	51	–	28	31	–
Norway	–	60	57	–	–	34	39	37
UK	–	57	–	54	26	26	–	35
US	56	55	–	54	25	32	–	35
Hungary	52	51	–	–	25	28	–	–
Finland	–	–	51	52	–	–	35	39

Source: Gershuny et al. (1994: Table 5.8).
Note: Paid work includes work breaks, unpaid work time excludes breaks, so men's total work time is overestimated.

changing? Table 5.4 reveals the limited extent to which men engage in unpaid work (contributing just over one-third of all unpaid work in most countries) and how this is failing to change significantly despite women's insertion into employment and working-age male unemployment. As Gershuny et al. (1994) conclude, men's proportion of unpaid work is growing but men's share of total work is decreasing in seven of the eight countries examined (i.e. Canada, France, the Netherlands, Norway, UK, USA, and Hungary). In the eighth, Finland, there is a small increase in men's proportion of total work. This reveals, in sum, that there is a process of gradual or 'lagged adaptation'. With women's increasing entry into paid employment, their total workload increases, but this is moderated, though not entirely offset, by the substitution of some male unpaid work (see Gershuny et al. 1994).[4] Overall, therefore, one can assume that neither the extent nor the character of women's participation in informal employment is undergoing radical change since they remain primarily responsible for domestic duties and as such, find work which either reflects their domestic roles (with the resultant sectoral divisions) or fits in with their domestic duties (resulting in participation in regular but part-time informal employment so that family needs can be met).

Having outlined these strong similarities across the advanced economies in the gender divisions of labour which configure the extent and character of informal employment undertaken by men and women, it needs to be high-lighted that this does not mean that the gender configuration of the informal labour market is everywhere the same. Despite the prevalence of such general tendencies, there is also evidence of cross-national, regional and local differences both in attitudes and behaviour concerning the gender division of domestic labour. Kiernan (1992) reports the results of a series of Euro-barometer surveys on domestic work (1975, 1978, 1983, 1987). Respondents were asked which of the following arrangements corresponded most closely to their idea of a family:

1 a family in which both husband and wife have equally absorbing work, and in which the household tasks and looking after the children are shared equally between husband and wife (egalitarian option);
2 a family in which the wife's work is less absorbing than the husband's, and in which she takes on more of the household tasks and looking after the children (middle option);
3 a family in which the husband only works and the wife runs the home (homemaker/breadwinner option).

This identified wide cross-national variations in these social norms and moral-ities concerning who is responsible for the domestic workload and whether women should engage in formal employment. One extreme is Denmark. Danish men and women were most likely of the EU citizens to prefer an egali-tarian model (51 per cent of men and 55 per cent of women), the most likely to eschew the traditional model (12 and 13 per cent of men and women respectively) and the responses of the genders were in accord. At the other extreme are Ireland, West Germany and Luxembourg. Here, the proportion of men and women preferring the traditional model is greater than those preferring the egalitarian model. For example, 34 per cent of German women and 29 per cent of German men prefer the traditional model compared with 27 per cent of women and 25 per cent of men preferring the equal roles model. Traditional attitudes are also strong in Belgium. When examining who usually took care of various domestic tasks, furthermore, Kiernan (1992) finds much the same rank order of nations according to the extent to which men participate in domestic work. The implication, therefore, is that these cross-national (and doubtless intranational) variations in the ideology and practice of the domestic division of labour will have an influence on the gender configuration of informal employment, particularly when coupled with the economic characteristics of the area which dictate the opportunities in the formal labour market afforded to women. Since no explicit research has been carried out in this regard, it is difficult to draw any hard and fast conclusions. However, it might be conjectured that since women's informal employment is strongly correlated with economic necessity, for relatively affluent populations, attitudes towards women's domestic and caring responsibilities will have little effect on their informal employment. More egalitarian attitudes and practices will lead to more formal and not more informal employment for women. Conversely, for poorer populations, it may well be the case that attitudes towards women working outside the home and men's responsibilities within the home will have an impact on the extent to which women's or men's labour is used in informal activity in order to allow the family to get by.[5]

In sum, it is not just the structure of the formal labour market that explains the gender configuration of informal employment. The gender division of domestic labour is also a major factor contributing to this gender configuration. Ideologies that view domestic work as women's responsibility narrow the types of informal employment available to them. However, it is not solely the way in which economic restructuring shapes, and is shaped by, social structures that configures the gendered nature and extent of informal employment in different places. These economic and social structures are also mediated by institutional factors.

Institutional characteristics

Institutional influences include the structure of labour laws, welfare benefit regulations and state interpretation and enforcement of these laws and regulations, as well as the nature of corporatist agreements. These all affect the gender configuration of informal employment either directly or indirectly. Indeed, it is such institutional factors (varying greatly between nation-states) that have contributed heavily to the ways in which women are integrated (or not) into the formal labour market. This, as we have discussed above, has an important effect on women's participation in informal employment.

For example, Cousins' (1994) comparison of the ways in which women's labour has been used to provide labour market flexibility in Britain and Spain is a case in point. In recent years, employers throughout the advanced economies have drawn on the hidden labour supply of married women who provide flexibility in the labour market since it is they who have been prepared to work in atypical jobs. However, in countries in which the state acts with a heavier hand in regulating the labour market (in theory, at least), a proportion of this flexibilisation has had to be undertaken using informal employment. Therefore, Cousins (1994) finds that in the UK, where labour laws have been deregulated to a greater extent than Spain, women's incorporation into the labour market has taken the form of formal part-time employment, whilst in Spain, it has tended to take the form of either temporary work or informal employment.

The state, moreover, can play a direct and active role in some circumstances in configuring women's informal employment. As Gringeri (1996) points out with regard to the US, the practices of the state in encouraging informal and formal home-based work in the rural mid-western states has acted to cement an understanding of home-based work as marginal employment for women. Using informal and formal homework as part of their economic development strategies in these areas, the local state has constructed such work as 'women's work' and reinforced the gender division of labour. Similar tendencies have been discerned in both Finland (Salmi 1996), Canada (Dagg 1996) and the UK (Tate 1996).

However, in other circumstances, the state has acted to improve women's working conditions rather than reinforce gender divisions of labour. In the Red regions of Italy, for example, the communist administrations have encouraged women's informal (and formal) employment and provided better services to make it possible. Nursery schools in Reggio Emilia and Modena, for example, are sufficient to absorb the entire demand for their services. Nevertheless, even here, women continue to be assigned primary (or sole) responsibility for both unpaid domestic work (Vinay 1985) and 'women's work' (e.g. child-care). This is perhaps because although many advanced economies often shift activities such as child-care in and out of the informal sphere, this has had little influence on men's participation. The result is that women exploit other women so as to liberate themselves but no renegotiation of men's role is undertaken (Briar 1997, Windebank 1996). State policy, therefore, often serves merely to reinforce such gender divisions.

Indeed, with the increase in women entering formal employment, there has been something of a boom in care services for the household, whether this is for child-care or domestic cleaning (Gregson and Lowe 1994). Such services are closely linked with informality. Some states, for example, France and Belgium, have taken notice of this trend and consequently have attempted to formalise these traditionally informal work arrangements by offering employers subsidies to their tax and social security contributions in respect of these workers. The jury is still out, however, on the success of these schemes in formalising the position of many women informal employees.

Conclusions

This chapter has shown that the gender configurations of informal employment reflect the gender divisions of labour more generally. Although women participate in informal employment to a lesser extent than men, when they do engage in such work it is more likely to be low-paid and exploitative informal work, whilst men engage in the higher-paid and more autonomous forms of such work. Women, moreover, tend to conduct informal employment on a regular but part-time basis due both to the constraints of their domestic roles and responsibilities and the fact that their motivation is economic, based upon the desire to generate extra income to help the family get by. For them therefore, the necessity is for regular income from such work. For men, on the other hand, informal employment is more irregular but full-time and is often undertaken for the purpose of earning spare cash for socialising and differentiating themselves from the domestic sphere and women. Examining who does such work, furthermore, it tends to be non-employed women but employed men. This is because women's informal employment is mostly in the realm of domestic services which remain inconspicuous to the outside observer and are perceived to be within the capability of all women, whilst men's informal employment tends to be home repair, maintenance and building work which is more conspicuous and is thought to require greater skills which the unemployed are not perceived to possess. The result, therefore, is a clear gender segmentation of the informal labour market both in terms of the work undertaken, motivations, pay and the types of men and women who undertake such work.

Consequently, just as women's participation in formal employment is of a different nature to that of men, their participation in informal employment is also different. Their domestic-orientated, exploitative and low-paid informal employment is predicated on the basis of their role as domestic workers and has to fit in with their domestic duties and responsibilities. If women find themselves in this type of informal employment, therefore, it is because they find themselves in this kind of work in formal employment, not because informal employment is comprised of this sort of work alone and these sorts of workers only. The problem is that women's identity as informal employees is more often than not rooted in traditional expectations and gender hierarchies. Their informal employment is closely associated with domestic work that diminishes its professional legitimacy not only at the macro-level in the hierarchy of professions but also in the perceptions of other family members. Therefore, there is no guarantee that earning an informal income will reduce gender inequalities, nor even that it will have a positive impact on a woman's relative power within the family.

Nevertheless, just as employment embodies conditions both of exploitation and opportunity for women, the same is true of informal employment. It provides the practical advantage of making it easier for women to juggle their various roles and responsibilities. At the same time, it creates the disadvantage of obscuring the perceived amount of time and energy women invest in increasing the wealth and well-being of the family. These two aspects of informal employment do not negate each other and are not mutually exclusive. There is no question that informal employment, especially for women, involves exploitation. Workers are excluded from welfare benefits, paid low wages and have greater job insecurity. However, such work also provides the possibility of earning income for many women who would otherwise not be able to do so. This advantage cannot be ignored. On balance, therefore, is informal employment exploitative or an opportunity? And can informal employment be reconstructed in such a way that the balance is tipped in favour of opportunities rather than exploitation? If so, how? It is to these issues, which are common to all marginalised groups who undertake such work, that we shall return in Part III of the book.

6

ETHNICITY, IMMIGRATION AND INFORMAL EMPLOYMENT

Introduction

The aim of this chapter is to evaluate the nature and extent of the engagement of ethnic minority and immigrant populations (whether legal or illegal) in informal employment. Is informal employment more common in ethnic minority and immigrant populations? Is the informal labour market composed in large part of such groups? Why do they engage in this work? Do they undertake informal employment out of choice or is it an activity of last resort due to their exclusion from the formal labour market? What sort of informal employment do they conduct? Is it always exploitative and low-paid? How does their informal employment differ from that of the dominant white population?

In order to answer these questions, this chapter examines first, the extent to which ethnic minority populations and immigrants (legal or illegal) participate in informal employment and second, the character of the informal employment undertaken by these different groups. Using a case study of Miami, we then provide a richer description of the variations in the experiences of different ethnic minority and immigrant groups *vis-à-vis* informal employment in order to highlight how their differing life and work histories impact on the paid informal activities which they undertake. Finally, and in order to explain such differences, we reveal how economic, social, institutional and environmental characteristics combine in varying ways amongst different groups and in specific places to produce particular types of engagement in informal employment.

Before commencing, however, it is important to outline the limited nature of the information available. First and foremost, the vast majority of the evidence concerning the extent to which ethnic minority and immigrant populations engage in informal employment and the type of paid informal activities which they conduct derives from the US. The European literature on informal employment, in comparison, has on the whole been racially blind. In major part, this is due to the contrasting conceptualisation of the problematic of informal employment in contemporary discourse in the two regions. In the US, informal employment is part and parcel of the problem of immigration whereas in the EU, it is perceived to be more related to unemployment. Here, therefore, and due to the limited data elsewhere, we rely mostly on these US studies to explore the level and character of the participation in informal employment of these populations. These US studies, nevertheless, possess some inherent shortcomings. The vast majority of them focus upon low-paid, labour-intensive, non-unionised and exploitative industrial sectors and occupations in poorer areas with high concentrations of ethnic minorities and immigrants, mostly in global cities (Fernandez-Kelly and Garcia 1989, Lin 1995, Portes 1994, Sassen 1989, Stepick 1989). The inevitable result is that they identify what they seek: that informal employment is closely associated with such groups and that these groups engage in organised forms of exploitative, low-paid, informal employment. Despite these shortcomings, however, they do provide some important insights into the magnitude and character of informal employment amongst ethnic minority and immigrant populations as long as they are read in the context of the conditions of their production.

Extent of ethnic minority and immigrant participation in informal employment

Do ethnic minority and immigrant groups more frequently participate in informal employment than their white or indigenous counterparts? Is the informal labour market heavily composed of these groups? To answer these questions, two different groups need to be distinguished, each likely to have contrasting experiences with regard to the informal labour market. On the one hand, there are ethnic minority populations and on the other hand, there are illegal immigrants. Given that different types of legal immigrant ultimately witness the same experiences as one of these two groups, they are here dealt with under these two headings. Naturalised legal immigrants to all intents and purposes, that is, witness similar experiences to second- or third-generation ethnic minority populations with regard to their informal employment. Legal non-naturalised immigrants, meanwhile, come to a nation to take up a particular job (which, by definition, will be formal employment to allow a work permit to be agreed). If they lose this job, then repatriation will normally ensue, although this depends on the specific national laws and regulations in place at any time concerning work and residence permits. Consequently, the choices open to this group are either to find alternative formal employment, return to their home nation or become illegal immigrants or workers. Each group is now considered in turn.

The participation of ethnic minority populations in informal employment

In order to set the scene for examining the participation of ethnic minorities in informal employment, we commence by analysing the economic marginalisation and socio-spatial distribution of ethnic minorities in the advanced economies. Much of the literature on racial inequalities highlights a deepening racial polarisation in North America, as evidenced by rising unemployment and poverty among urban African-Americans (Galster 1991, Kasarda 1989, Massey and Denton 1993, Oliver and Shapiro 1997). In Britain in particular and Europe more generally, similar tendencies are to be found. Ethnic minorities have lower incomes, experience a disproportionately high level of unemployment, have a lower proportion of men in full-time work and a higher representation in semi- and unskilled manual jobs (Bhavnani 1994, Jones 1993, Modood and Berthoud 1997, Owen 1994).

For ethnic minority women, moreover, the situation is further compounded due to the 'double disadvantage' of both racial and gender discrimination in the formal labour market (Amott 1992). Their earnings are less on average than those of white women and ethnic minority men; they are more likely to work in the low-wage secondary segment of the labour force (Johnson-Anumonwo et al. 1994) and are overrepresented in service occupations like domestic service, clerical work and health aid and in certain manufacturing occupations such as textile machine work and assembly (Amott and Matthaei 1991, Woody 1992). Of course, not all ethnic minority women are confined to such peripheral jobs. The share of Chinese, Japanese and Filipino-American women in managerial and professional jobs in the US is equal to or higher than their share of the labour force in general but they are also overrepresented in low-paid manufacturing and family labour jobs (Amott and Matthaei 1991). Therefore, social polarisation in general is exaggerated amongst these East Asian women. In Europe, meanwhile, there is evidence that the gendered racial polarisation witnessed in the US is being repeated. In Britain, for instance, all recent analyses of women in the labour market reveal that ethnic minority women are more likely to be unemployed, overrepresented in areas of occupational decline and underrepresented in growth sectors, working longer hours, in poorer conditions and for lower pay than white women (Roberts 1994).

Inextricably interrelated to such labour market disadvantages are the spatial distribution patterns of ethnic minorities in the contemporary advanced economies. The segregation of minorities in poor inner-city neighbourhoods coupled with suburban job growth are highly visible features of the US metropolitan landscape (Farley and Frey 1994, Galster 1991, Jackson 1995, Massey and Denton 1993). This has led many geographers to highlight the 'spatial mismatch' between ethnic minorities and employment opportunities (Fainstein 1996, Holzer 1991, Kasarda 1989, McLafferty and Preston 1996, Wyly 1996).[1]

What are the implications of this economic and socio-spatial segmentation of ethnic minorities for informal employment? Do ethnic minorities in general and especially those concentrated in areas of high unemployment and poverty, engage in more informal employment than other groups in society? The vast majority of the US studies do indeed identify that informal employment is closely associated with such groups (Fernandez-Kelly and Garcia 1989, Lin 1995, Portes 1994, Sassen 1989, Stepick 1989). Most of these studies, nevertheless, focus upon sectors and/or occupations in areas with high concentrations of ethnic minority populations, mostly in global cities. The problem is that although it is clear that ethnic minorities in general and ethnic minority women in particular dominate in such sectors and occupations, this does not mean that they universally constitute the bulk of either the informal labour force in general or even the peripheral informal workforce more particularly.

Indeed, where ethnicity has been considered as a variable alongside others in alternative geographical contexts, such as in rural Pennsylvania (Jensen *et al.* 1995), no strong correlation has been identified between ethnic minorities and participation in informal employment. This indicates, therefore, that the informal labour market appears to be configured differently in different areas so far as ethnicity is concerned. It also indicates that if ethnic minorities are more likely to work informally in some parts of global cities, this does not mean that the vast bulk of the informal labour force in the advanced economies as a whole are ethnic minorities, as reading the US literature outside its context might imply. Informal employment, it seems, is not exclusively concerned with the economic adjustment of immigrant groups, internal migrants or domestic ethnic minorities. Although this might be the case in some localities and neighbourhoods (see Portes and Sassen-Koob 1987, Portes 1994), it is not universally the case.

The participation of illegal immigrants in informal employment

Superficially, there appear to be few doubts that the participation of illegal immigrants in informal employment is very high, possibly the highest of any socio-economic group in the advanced economies. The reason is supposedly simple. Excluded from formal employment as well as social security benefits due to their illegal status, such immigrants have little choice open to them but to engage in informal employment as a means of survival. However, in reality, they do have other options. They can use falsified or other people's documents (e.g. national insurance or social security numbers) to gain access to either welfare benefits or employment without their employers' collusion. Alternatively, they can work as an employee with the consent of the employer who pays taxes for the worker, with the worker remaining undetected as an illegal immigrant due to the imperfections in, and lack of co-ordination between the various regulatory institutions of the state. Indeed, the Internal Revenue Service (IRS) in the US has estimated that as many as 88 per cent of illegal immigrants may be paying taxes (cited in Mattera 1985). It is therefore not the case that all illegal immigrants who work are engaged in informal employment. Although obviously engaging to a greater extent in informal activity than other groups, the picture is much more complex than is sometimes thought.

Quite how many illegal immigrants work in informal employment, therefore, is open to debate. The EU estimate that there were over 600,000 illegal migrant workers in the Community in the mid-1970s (European Commission

1974) whilst De Grazia (1984) suggests the number may have grown with estimates of 300–400,000 in France in 1980 and 600–700,000 in Italy in 1981. Whether this can be taken as evidence of growth, however, is doubtful because there are wide fluctuations in estimates of the number of illegal immigrant workers. For example, in the US, Witte (1987: 73) reports IRS estimates of the size of the illegal alien population in 1979 as 3.5–6 million, of whom 65 per cent were employed, whilst Briggs (1984), for the same year, presents evidence to suggest that there were about 1.5 million illegal aliens. It is without doubt, however, that the magnitude of illegal immigration has been greater in the US than Europe. The land border of the USA with Mexico is long and difficult to patrol and the contrast in living standards in the two nations, as well as with other Central and South American countries, has led to considerable inflows of illegal immigrants (see Espenshade 1995a and b, Thomas 1992).

Although such inflows are perhaps lower in the advanced economies of Europe, this is not to say that illegal immigration is unproblematic. Take, for example, Italy. For many decades, this nation provided large numbers of emigrants but there were few immigrants (Mattera 1980, Willatt 1982). In 1978, however, the research organisation, CENSIS, estimated that there were 280–400,000 illegal foreign workers, mainly from northern Africa and ex-Yugoslavia, concentrated in Rome and the large industrial cities of the north. By 1984, however, the Italian national statistical office, ISTAT, estimated that there were half a million foreigners working illegally, whilst Mingione (1991) put the figure at anywhere between 300,000 and 1 million. Italy, however, is not alone in Europe in witnessing high numbers of illegal foreign workers. In 1987 in Greece, there were an estimated 150,000 non-declared foreigners working illegally (excluding foreign sailors working in the large Greek merchant fleet), many employed in restaurants, bars and discos. These, however, are not only composed of non-European but also young people from northern EU nations who work informally during the summer season for low pay and no security (Hadjimichalis and Vaiou 1989). Indeed, similar tendencies are identified throughout southern Europe by Hadjimichalis and Vaiou (1989) where jobs that involve hard manual work and low pay have been increasingly taken up by ethnic minorities and/or migrants from Third World or socialist countries. Such tendencies have also been identified in northern EU nations such as France (Moulier-Boutang 1991) and Germany (Jones 1994). Indeed, with the collapse of many of the socialist economies of central and eastern Europe, such problems have probably increased during the 1990s.

The important point here, however, is that even if the vast majority of illegal immigrants do engage in informal employment, this does not mean that all informal workers are illegal immigrants. As Wuddalamy (1991: 91) argues in France: 'Immigrant clandestinity . . . only covers fragments of the underground economy.' So, given these findings on the extent of participation of ethnic minorities and illegal immigrants in informal employment, attention now turns towards the nature of the informal employment that they undertake.

Nature of participation of ethnic minorities and immigrants in informal employment

Here we examine the various types of informal employment engaged in by ethnic minorities and immigrants compared with the dominant white population, along with their contrasting motivations and wage rates.

Ethnicity, immigration and types of informal employment undertaken

Examining the character of the informal employment undertaken by ethnic minorities and immigrants (whether legal or illegal), the overwhelming finding is that they engage in peripheral exploitative forms of organised informal employment rather than the core more rewarding autonomous types of such work (Lin 1995, Martino 1981, Mingione 1991, Mingione and Magatti 1995, Moulier-Boutang 1991, Phizacklea 1990, Portes 1994, Portes and Stepick 1993, Pugliese 1994, Sassen 1989, Stepick 1989, Zlolinski 1994). Therefore, the informal labour market mirrors many of the racial inequalities prevalent in the formal labour market. As Moulier-Boutang (1991: 119) puts it, 'all large world labour markets are hierarchical along ethnic lines: foreign workers and different nationalities in different legal positions occupy the posts at the bottom of the scale'.

This racial segmentation of the informal labour market is further cross-cut by gender so far as the sectoral and occupational division of informal employment is concerned. Ethnic minority and immigrant women find themselves confined to those occupations and sectors that also reflect their domestic roles and responsibilities or their domestic constraints. As such, they are overrepresented in specific sectors and occupations such as paid domestic service (Mingione 1991) and manufacturing homework throughout the advanced economies (Boris and Prugl 1996, Phizacklea 1990, Sassen 1989). Ethnic minority and immigrant men, meanwhile, also find themselves heavily concentrated in specific sectors and occupations, such as in the construction industry. In New York, Sassen (1989) finds that with the growth in sub-contracting to informal small construction firms, much of the informal labour used is Hispanic. In retailing, in addition, not only do a large number of men engage in autonomous informal work such as street vending (Portes 1994, Stauffer 1995) but increasingly, this has become dominated by ethnic populations and immigrants especially in ethnic neighbourhoods where they sell goods imported from their home country to people of their own ethnicity. In a sense, therefore, ethnic minority and immigrant men are more likely than women to be engaged in relatively autonomous informal employment, although this is further confused in reality by the fact that street vendors may be working for somebody else whilst women homeworkers may be employed on a self-employed piece-work basis, although they will usually have only one employer to whom they supply the goods produced (Boris 1996, Phizacklea and Wolkowitz 1995).

The nature of informal employment undertaken by immigrants and ethnic minorities, however, is mediated not only by gender but also by how immigration takes place and by the length of time that immigrant groups have been resident in their host nation. Starting with the way in which immigration occurs, the nature of informal employment will be influenced by whether such immigration is legal or illegal, which itself will be shaped by the structure of informal employment in the host nation. Sometimes, for instance, it is women who arrive first, followed by men, such as in the case of Mauritian immigration into France (Wuddalamy 1991). Here, women entered first by finding jobs through their co-ethnic contacts in domestic service where detection by the French authorities was difficult. Husbands then followed, often finding informal employment in the garment trade in the Sentier district of Paris using their ethnic, national and religious bonds to seek out ethnic employers.

The character of the informal employment undertaken is also the result of the length of time that the immigrant group has been settled. This is clearly displayed by Sassen (1989) in the New York apparel industry where a highly vertical production structure has developed, with recent immigrant workers (mostly female) at the bottom; immigrant male contractors in the middle; older ethnic groups, predominantly Jews, as manufacturers and wholesalers; and large retail chains and stores at the top. The older the immigrant group, therefore, the higher up this hierarchy it finds itself (Lin 1995, Waldinger 1986). Many informal firms established by older ethnic groups, moreover, slowly become formal companies over time, as Sassen (1989) found in the electronics industry in New York. The result is that different ethnic minority and immigrant populations, as well as different sub-groups within these populations, engage in varying types of informal employment. Not all ethnic minority and immigrant groups conduct exploitative, low-paid, organised forms of informal employment and a certain amount of social mobility appears to be possible for some ethnic groups. Many informal employers, for instance, are from ethnic minorities (Portes 1994, Sassen 1989). They may be from older immigrant communities who are more established and make use of recent immigrant groups as a source of cheap informal labour in their businesses, or they may be wealthier immigrants from the same group. In ethnic enclaves, for example, many ethnic minorities and immigrants are employed informally by employers of the same ethnicity (Lim 1993, Pugliese 1994, Sassen 1989, Portes and Stepick 1993).

Ethnicity, immigration and motivations for undertaking informal employment

So far as illegal immigrants are concerned, it might be assumed that their motivations for engaging in informal employment are clear-cut. Excluded from formal employment, they have little choice but to engage in informal employment in order to generate the resources required for survival in their host nation. However, and as indicated above, such immigrants do have other options. This means that they are not perhaps so constrained to informal employment as has been assumed. There is a need, therefore, for much greater research on the motivations of illegal immigrants for engaging in informal employment. Is it a matter of choice? What part do social networks and motivations in the form of common ethnic bonds play in their decisions? What are the constraints on their choices? How does this vary between different groups of illegal immigrants and between different areas? How do state regulations and laws concerning immigration affect their choice? These are just some of the questions which will need to be answered if one is to more fully understand why illegal immigrants engage in informal employment.[2]

For ethnic minorities, meanwhile, it seems likely that their motivations for engaging in informal employment will be similar to those of other unemployed and low-income groups (see chapter 4). That is, those who are unemployed or poor tend to engage in informal employment on a more involuntary basis due to their lack of other options, compared with affluent groups who conduct autonomous informal work on a more voluntary basis. This is influenced by a range of additional factors such as the cultural traditions, norms and moralities concerning informal employment of both the ethnic group and the area in which they live more generally. Take, for example, a study of the contrasting motivations of white and black groups receiving informal payments in US College Football. Sack (1991) shows that black athletes were somewhat more likely to see nothing wrong with violating the rules since there is less acceptance or 'ownership' of state rules and regulations amongst some ethnic minorities which leads to less concern about transgressing them. Although this is a dangerous argument, for it implies that ethnic minorities and immigrants are more lawless, other studies of informal employment in different contexts, such as amongst Catholics in West Belfast, have come to much the same conclusion (Leonard 1994). The salient point here, therefore, is that economic necessity alone is insufficient to explain why some members of ethnic minorities and immigrants engage in informal employment. The issue of their motivations is far more complex.

Ethnic minorities, immigrants and informal wage rates

The overwhelming finding is that both ethnic minorities and illegal immigrants engage in the more exploitative kinds of informal employment compared with the dominant white groups and are thus lower paid both in terms of annual earnings and average hourly rates, although significant differences are identified between different ethnic minority and immigrant groups which are cross-cut not only by gender but also by the amount of time such groups have been inserted in the host nation (Lin 1995, Portes 1994, Portes and Stepick 1993, Pugliese 1994, Sassen 1989, Stepick 1989).

To move beyond such generalised statements about wage rates, however, is difficult. All that exists are idiographic findings of specific sectors in particular localities. To take just one example, Costes (1991) studies illegal sellers in the Paris underground and identifies a racially-based hierarchy reflected in both status and wage rates. Of the 300 sellers identified in 1989, the 11 per cent who were French and were mostly autonomous informal workers had turnovers estimated at 180,000 FF per annum. The Tunisian sellers, meanwhile, who were unemployed immigrants with residence permits, and working as informal employees selling fruit for a wholesaler, were paid 200–300 FF per day for their work. The remainder, mostly illegal immigrants from the Indian sub-continent, who use their co-ethnic networks to receive goods to sell and find a pitch in the Metro, earned considerably less than the former two groups. However, to apply such findings more generally would be perhaps unjustified, although they do tally with more general understandings of the plight of different ethnic and immigrant groups.

In conclusion, although ethnic minorities and illegal immigrants do not form the vast bulk of the informal labour force, where they do engage in such work, it tends to be mostly of the peripheral exploitative kind; their motivations are mainly economic and they receive low pay for the work they carry out. Nevertheless, although this is the dominant finding, important exceptions to this rule can be identified. Ethnic minorities and immigrants (whether legal or illegal) do not always occupy the most peripheral positions in the informal labour market. Rather, different ethnic minority and immigrant groups are to be found in different positions in the hierarchy of particular local informal labour markets and these positions can change over time as some groups progress up the hierarchy of paid informal activities and even formalise them. Informal employment, therefore, has a hierarchy of its own which mirrors the formal labour market so far as ethnicity and immigrant populations are concerned. To provide a richer description of the more complex ways various ethnic minority and immigrant groups relate to the informal labour market in a particular place, we now take a case study of Miami for investigation.

The case of immigrants in Miami

The case of Miami is similar to elsewhere in the sense that the magnitude of informal employment, especially the difference between the extent of participation of the white population and various ethnic minorities and immigrants, is not known. A survey of Cuban refugees and Haitians, however, does provide some evidence of the extent of the participation in informal employment of these two immigrant groups. This finds that about third of those who reported themselves employed in 1983 were working informally through cash payments, not paying benefits or receiving hourly wages below the legal minimum (Portes and Stepick 1993). The vast bulk of this informal employment, moreover, was low-paid, organised, informal work in the retail, restaurant, apparel and construction industries (Portes and Sassen-Koob 1987).

Stepick (1989) and Portes and Stepick (1993) examine the contrasting experiences of the successive waves of Cuban, Nicaraguan and Haitian immigration into the US. Although Washington welcomed the Cubans who had their passage paid, were automatically granted residence in the country and received numerous benefits, the federal government classified most Nicaraguans and Haitians in contrast, as illegal aliens and actively tried to keep them out. Nevertheless, and despite Cubans being legal and Nicaraguans and Haitians illegal immigrants, the pattern of their immigration was very similar. The Nicaraguan exodus, like the Cuban one before it, converged in successive stages on Miami, beginning with the elites, then incorporating the professional and middle classes and lastly the working class.

For each immigrant group, the ways in which they inserted into, and used, informal employment have differed. For the Cubans, the exodus of the middle class following the Cuban revolution instigated the growth of informal enterprise in a number of sectors in Miami. Employing their past business experience as well as the solidarity fostered by their common past, informal entrepreneurs expanded their operations and moved them into the formal sphere, using later waves of working-class Cubans as a source of labour. These enterprises, moreover, were fully integrated with the wider Miami economy in the sense that goods and services were bought and sold outside the emerging ethnic enclave. The relationship between employers and employees was such that although employees voluntarily submitted to high levels of exploitation, they received in return assistance in subsequently establishing and maintaining their own business. It was not solely the Cuban middle class, however, which made use of this new source of cheap and plentiful labour in the form of the Cuban working class. Older immigrant groups, such as New York Jewish manufacturers, escaped unionisation drives in New York by relocating to Miami to employ Cuban refugee women arriving during the 1960s.

As Cuban immigration ground to a halt, however, these enterprises (both formal and informal) in the Cuban ethnic enclave started to witness labour shortages. Fortunately, a ready source of new labour became available in the form of Nicaraguan immigrants. Take, for example, the construction industry. The Latin Builders Association – Cuban-owned firms set up in the late 1960s and early 1970s and grouped in a guild – confronted serious labour supply problems in the mid-1980s. Cuban immigration had virtually stopped and the builders were reluctant to employ union-prone labour. The Central Americans in the form of Nicaraguans filled the gap. Their need for work and ideological affinity with their employers meant that they became the preferred workers in the Miami building trades. The arrival of the Nicaraguan working class thus had a significant impact on the Miami economy. All employers of low-skilled labour benefited, especially garment manufacturers and house builders, farmers and middle-class families searching for domestic helpers. More specifically, the new immigrants helped the Cuban enclave avert a serious problem, expanding both their market and labour pool. The Nicaraguans did have the advantage, however, of arriving after the Cuban enclave had been established and so had a powerful ally. This was not solely due to their Latinness, but also due to political ideology – the common circumstances of militant opposition to an extreme-left regime.

For the Haitians, it was a different story. As with the broader Miami economy in general, the ethnic enclave showed a reluctance to admit Haitian immigrants into jobs. So, despite being an even cheaper and more pliable source of labour than the Nicaraguans, they were not incorporated into the formal and informal ethnic enclave workforce. Instead, solidarity was shown with the Nicaraguans. The result, when coupled with US society's rejection of their presence, was that the Haitians set up their own more isolated enclave which consisted mostly of intra-Haitian trading. As such, the Haitians created their own economy (composed of both formal and informal elements) in which the informal element closely resembles the original descriptions of the phenomenon in Third World cities: casual self-employment isolated from the broader market and used primarily as a survival mechanism. This reflects the different economic conditions prevailing when the Haitians arrived in Miami and the contrasting institutional attitudes and behaviour towards them as well as the important social effects of a hierarchy of racial prejudice operating in Miami that ranges from whites through Hispanics to the black immigrants from Haiti.

Different waves of immigration, therefore, resulted in various configurations of informal employment in Miami. This is not to say that exactly the same processes are at work elsewhere, even amongst the same immigrant groups. Fernandez-Kelly and Garcia (1989) examine homeworking amongst Hispanic women in the Los Angeles and Miami garment industries. In Miami, they discover that the existence of an ethnic enclave formed by Cuban entrepreneurs enabled women from the same families and community to use homework as a strategy for improving earnings and for reconciling cultural and economic demands. In Los Angeles, in contrast, Mexican immigrants have not formed an economic enclave. Instead, most workers have entered the labour force in a highly atomised manner at the mercy of market forces entirely beyond their control. As Fernandez-Kelly and Garcia (1989: 263) thus conclude: 'involvement in informal production can have entirely dissimilar meanings depending on the type of incorporation into the labour market.' This reinforces the notion of an informal labour market that is both locally specific and socially differentiated. How to explain these socio-spatial variations in the extent and nature of informal employment amongst different ethnic minority and immigrant groups is the issue to which attention now turns.

Explaining variations in the extent and nature of informal employment amongst ethnic minorities and immigrants

To explain variations in the magnitude and character of informal employment undertaken by ethnic minorities and immigrants both across and amongst social groups and space, we here draw upon the same theorisation employed throughout this book. Such variations are seen as a product of the way in which a range of economic, social, institutional and environmental characteristics combine in different ways in different areas to produce particular local outcomes. Here, therefore, we explore some of the important characteristics which influence how the informal labour market is experienced by different ethnic minorities and immigrants in contrasting areas.

Economic characteristics

Starting with the economic characteristics, two of the key factors which will shape the experience of ethnic minority and immigrant groups with regard to informal employment are the configuration of the local formal labour market and the level of unemployment in the area to which they move. The nature of the local formal labour market, and the structure and power of ethnic enclaves within these labour markets, for example, is an important determinant of whether and how these groups become embedded in informal employment in a particular locality. As we have already discussed, the comparative study of informal industrial homeworking amongst Hispanic women in the garment industries of Miami and Los Angeles by Fernandez-Kelly and Garcia (1989) clearly reveals quite different local outcomes due to the contrasting structure of the local labour market in the two areas. In Los Angeles, such work took on a highly proletarianised form, associated with the hyper-exploitation of successive waves of Mexican immigrants. For the women involved, many of whom were undocumented workers, homework was a strategy of last resort; it raised the household above the poverty line and was pursued by women in particularly vulnerable positions in the labour market. In contrast, in Miami, homework had a different underlying logic. The presence of a Cuban entrepreneurial class, and the opportunities this gave women to work within their (albeit patriarchally organised) community, afforded a degree of shelter from the market forces faced by Mexican homeworkers in Los Angeles. The same process, therefore, had different outcomes in these contrasting labour markets.

The level of unemployment, moreover, is an important determinant of how and if ethnic minorities and immigrants immerse themselves into informal employment. As unemployment increases, for example, a common trend is for the dominant white population to replace ethnic minorities and immigrants not only in the formal labour force but also in informal employment. Jones (1994) shows this to be the case in Germany and documents the ways in which such groups have been squeezed first out of the formal labour market and then increasingly out of informal employment. Of course, whether this will occur in every locality, region and nation is also dependent upon a range of other factors, such as the nature of the welfare system and the rules concerning immigration, to name but two. In nations with comprehensive and universal welfare systems, for example, the degree to which illegal immigrants are forced out of informal employment by the dominant white population will be much lower than in those nations lacking such a comprehensive welfare system. Indeed, whether the social security system is universal or insurance-based will also affect the extent to which immigrant workers (as well as the native population) who have not spent all of their working lives in the host nation will engage in informal employment. The ways in which economic characteristics influence the nature and extent of the participation of these groups in informal employment, therefore, is mediated by institutional characteristics.

Institutional characteristics

The principal institutional characteristics which shape the nature and extent of the participation of immigrants and ethnic minorities in informal employment include the nature of the welfare and immigration laws, and the extent to which they are enforced, as well as corporatist agreements. We have already briefly discussed the ways in which the configuration of the welfare system might influence the relationship between informal employment and ethnic minorities and immigrants. Here, therefore, let us examine the other principal institutional factor in the form of immigration policy.

Immigration policy varies significantly between nations and these variations can have major impacts on the extent to which ethnic minorities, but more particularly, immigrants, engage in informal employment. In nations where residence permits are closely tied to work permits, the number of legal immigrants available to work informally will be much lower than in nations displaying a looser connection between work and residence permits, since immigrants may enter the country legally without the right to employment. One particular instance of legal residence without the right to work is that of family reunion where spouses and dependants of existing immigrants are allowed to live in the country but are not allowed to be employed. Levels of informal employment amongst immigrants will also be higher in those countries where access to welfare benefits is denied and there is a long waiting period before access to the labour market is granted. The state, therefore, through its immigration policies, can play an active role in configuring the nature and size of the informal labour market.

Social characteristics

Besides economic and institutional characteristics, the configuration of informal employment *vis-à-vis* ethnic minorities and immigrants is also shaped by social characteristics. These include the nature and extent of social networks in the host nation, the traditions, cultural norms and moralities of the ethnic group as well as whether and how these are transferred to the host nation by the immigrant group and/or ethnic minority, and the socio-economic mix both of the immigrant group and the area under consideration.

Take, for example, the influence on informal employment of kinship and community relationships in ethnic minority and immigrant communities as well as the socio-economic mix of these communities. Where an ethnic group is characterised by a broad socio-economic mix and strong social networks, one may witness the development of ethnic enclaves and the result might well be that members of this group are employed (either formally or informally) by their compatriots, as in the case of Cubans in Miami. Indeed, where one new immigrant group enters a host nation already inhabited by another minority from a similar situation, as in the case of Nicaraguans in the Cuban enclave in Miami, and in favourable labour market conditions, interlinkages between the two communities are likely to develop. However, where 'otherness' rather than similarity is perceived, as in the case of the black Haitians' arrival in Miami, the outcome may be exclusion not only from the dominant white 'enclave' but also from the existing ethnic enclaves and their opportunities for both formal and informal employment.

Whether and how informal practices are transposed from the home to the host nation amongst immigrant groups can also lead to variations amongst different immigrant and ethnic groups. This is clearly displayed in a study by Herbert and Kempson (1996) of informal credit systems amongst different immigrant groups in the UK. They discover distinct differences between ethnic minorities in how credit needs are met. Comparing a Bangladeshi and Pakistani community in Oldham and an African-Caribbean community in Brixton, they find that these three groups were remarkably similar, both in their level of borrowing and in their reasons for doing so, to other low-income households in general. However, there were significant differences in the sources of credit used. The African-Caribbean community used informal community-based credit schemes known as 'partners' (i.e. a rotating savings and credit association in which a group of people regularly deposit a fixed amount of money into a central fund which is then distributed to members in a pre-arranged order). The Pakistani community had a similar informal scheme known as 'bond-committees', but there was a notable absence of such informal credit schemes in the Bangladeshi community. This is because such schemes are only set up by communities where they exist in their country of origin and they are only applicable for people in regular work. In the Bangladeshi community, suffering high unemployment and without such a system in their home country, this type of scheme was not used.

However, social characteristics are not the only factor which interacts with economic and institutional characteristics to produce specific local configurations of informal employment.

Environmental characteristics

Environmental characteristics are the final set of factors which combine with economic, institutional and social factors to shape the configuration of informal employment amongst ethnic and immigrant groups in particular areas. Here, factors such as whether these groups live in dense spatial concentrations or are widely dispersed are important. The popular prejudice is that those groups living in tight ethnic enclaves are more likely to engage in informal employment than those living in more dispersed settings. However, there is little evidence that those living in ethnic enclaves have greater opportunities for conducting informal employment than those living in areas with a more diverse socio-economic mix. No research has so far been conducted to explore whether this is the case or not. For the moment, therefore, as with the unemployed, we must assume that in certain circumstances it will be those in areas with a more diverse socio-economic mix who will be more likely to conduct informal employment.

The implication is that less emphasis needs to be put on ethnic minorities and immigrants living in ethnic enclaves when studying informal employment. Just as Fainstein (1996) cautions against too much emphasis being put on the black population living in ghettos when studying poverty, due to the fact that two-thirds of the poor in the US (using the government's definition) are white and living elsewhere, the same applies to considering only ethnic minorities and immigrants in ethnic enclaves when studying informal employment.

Conclusions

This chapter has shown that ethnic minorities and immigrants, similar to other marginalised and disadvantaged groups (e.g. the unemployed and women), do not constitute the vast bulk of the informal labour force and neither is their participation necessarily higher than that of other groups. However, when they do engage in informal employment, the ethnic segmentation of the informal labour market is revealed to reflect closely the racial inequalities prevalent in formal employment. A core informal workforce has been shown to exist, composed mostly of the dominant white population, whilst ethnic minority and migrant groups are more likely to be found in the peripheral informal workforce conducting exploitative forms of low-paid informal employment. Nevertheless, although this is the dominant finding, important exceptions to this rule are identified. This chapter has revealed how different ethnic minority and immigrant populations as well as sub-groups within each population often take up different positions in the hierarchy of particular local informal labour markets. Equally, it has demonstrated how these positions change over time as some groups progress up the hierarchy of the informal labour market and even formalise their activities. However, just as in the formal labour market, such exceptions should not hide the fact that the vast majority of ethnic minorities and immigrants find themselves in peripheral jobs; the same is true of the informal labour market.

To explain the nature and extent of informal employment with regard to ethnicity and immigration, therefore, this chapter has drawn together the various economic, institutional, social and environmental characteristics that combine together in various ways in different places to produce particular configurations of informal employment. In so doing, it has revealed that explanations for the socio-spatial divisions of informal labour so far as ethnic minorities and immigrants are concerned are similar to those utilised earlier to explain gender and employment status differentials.

7

SPATIAL DIVISIONS IN
INFORMAL EMPLOYMENT

Introduction

Based on the assumption that informal employment is principally a survival strategy used by marginalised groups such as the unemployed, ethnic minorities and immigrant groups, the popular prejudice is that such employment is more prominent in locations where these populations are concentrated. Elkin and McLaren (1991: 217), for example, talk of 'disadvantaged localities where informal employment is often very important', whilst Haughton et al. (1993: 33) assert that 'the distribution of informal work . . . may well be especially important in areas of high unemployment, in part acting as a palliative, in part merely recycling people between employment and unemployment and possibly reducing official unemployment statistics'. For many commentators, therefore, informal employment is most likely to be found in those areas where such marginalised populations are concentrated: deprived inner-city localities (Blair and Endres 1994, Elkin and McLaren 1991, Haughton et al. 1993, Robson 1988) and poorer peripheral regions (Button 1984, Hadjimichalis and Vaiou 1989).

Drawing upon the empirical evidence from studies conducted throughout the advanced economies, the aim of this chapter is to evaluate critically the validity of this assumption about the geography of informal employment. To do this, first, we analyse the notion that informal employment is concentrated in areas dominated by marginalised populations. This reveals that poverty alone cannot be used to explain the spatial configurations of informal employment. Second, therefore, and drawing upon the regulators of informal employment already discussed in some depth in chapter 3, we present a tentative typology of the characteristics of localities which undertake high or low levels of mostly autonomous or exploitative informal employment. The outcome is a heuristic framework for better understanding the spatial divisions of informal employment.

To what extent is informal employment concentrated in deprived areas?

To evaluate the degree to which informal employment is clustered in deprived areas, we draw upon evidence, first, from indirect macro-economic estimates of the size of the 'underground economy'[1] and second from various direct studies of informal employment in specific localities. The former allow preliminary and tentative cross-national comparisons to be made whilst the latter enable regional and local variations in the magnitude and character of informal employment to be understood more fully.

International variations in informal employment

So far as is known, there are currently no surveys that use direct methods to estimate international variations in informal employment. Instead, one has to rely on indirect methods that look for evidence of informal employment in macro-economic data that have been collected and/or constructed for other purposes. As discussed in chapter 2, these techniques are severely limited in their accuracy. Such measurements cannot differentiate between informal employment and all other forms of criminal activity and transactions; they are based on unproved assumptions; and different macro-economic approaches produce dramatically different results in terms of the percentage of Gross National Product (GNP) that informal employment supposedly represents even within the same country at the same time. None the less, such indirect macro-economic estimates are the only source of information available on cross-national variations in the propensity to undertake informal employment. Indeed, despite their shortcomings, these estimates do provide a fairly consistent picture of the relative size of informal employment in different countries. As Barthelemy (1991) notes, monetary approaches always result in a similar hierarchical order in terms of the cross-national variations in the size of informal employment, suggesting that significant cross-national differences do exist and that these patterns can be detected even if the volume of this activity cannot be measured precisely using such techniques. As Table 7.1 shows, such macro-economic estimates suggest that the level of informal employment is higher in poorer countries and/or those with weaker welfare states than in more affluent nations and/or countries with more comprehensive welfare systems, always assuming that the proportions of informal employment and criminal activity remain stable across different countries.

Take, for example, the nations that comprise the EU. Summarising the findings of the EC Official Expert final synthesis report, *Underground Economy and Irregular Forms of Employment* (European Commission 1990), the European Commission (1991: 130) concludes that in most northern parts of the EU, informal employment is some 5 per cent of the level of declared work or less,

Table 7.1 Estimates of the magnitude of informal employment obtained through indirect methods, as a percentage of GDP

	Smallest	Highest	Average estimate
Ireland	0.5	7.2	3.9
Austria	2.1	6.2	4.2
Norway	1.3	9.0	5.5
Britain	1.0	34.3	6.8
Australia	3.5	13.4	8.4
Germany	3.4	15.0	8.7
Netherlands	9.6	9.6	9.6
Denmark	6.0	12.4	10.1
Sweden	4.5	14.1	10.1
Canada	1.2	29.4	10.7
Belgium	2.1	20.8	10.9
Spain	1.0	22.9	11.1
USA	5.0	28.0	11.3
France	6.0	23.2	11.4
Portugal	11.2	20.0	15.6
Italy	7.5	30.1	17.4
Greece	28.6	30.2	29.4

Sources: Derived from Dallago (1991: Table 2.1), Pestieau (1989) and European Commission (1990).

somewhat more in France and Belgium, and possibly reaching 10–20 per cent in the southern EU countries. Hence, these indirect methods appear to validate the suggestion that less affluent nations and/or nations with weaker welfare states (e.g. Greece, Portugal) engage in greater amounts of informal employment than richer nations and/or countries with more comprehensive welfare systems (e.g. Germany, Norway).

However, although such indirect estimates offer an approximate guide to the pattern of cross-national differences in informal employment, this is the limit of their usefulness in addressing the question of spatial divisions in informal employment in the advanced economies. These studies reveal nothing about local or regional variations in informal employment. Nor do they aid explanation. Instead, these studies which scrutinise spatial disparities merely in a cross-national sense tend to lead those analysing the results to explanations which are nationally-specific, for example, the tax burden or labour law, when these factors taken in isolation are not necessarily able to account for the existence of informal employment. To see this, one needs to go no further than examining the large variations in the character and degree of informal employment which exist within particular nation-states.

Regional and local variations in informal employment

A barrage of direct empirical studies of informal employment conducted in specific locations in countries across the advanced economies are reviewed here in order to draw conclusions concerning the relationship between poverty and the nature and volume of informal employment undertaken by localities and regions. First, we appraise the relationship between poverty and the magnitude of informal employment and second, we evaluate how poverty influences the nature of informal employment in localities and regions.

Reviewing the vast array of locality studies, the dominant finding is that even if the results derived from indirect methods are correct, that is, that poorer nations engage in more informal employment than affluent nations, it is incorrect that such employment is higher in the poorest localities and regions of individual countries. Quite the opposite. In the Netherlands, for example, a study of six localities by Van Geuns *et al.* (1987) demonstrates that the higher the rate of unemployment in an area, the lower is the level of informal employment. This finding is replicated in Britain. A survey of eight localities by Bunker and Dewberry (1984) discovers that those areas with the highest unemployment rates undertake relatively little informal employment. In France, meanwhile, studies in both Orly-Choisy (Barthe 1985) and Lille (Foudi *et al.* 1982) identify poverty black spots in which the unemployed cannot escape from their deprivation through informal employment because not only is less money available to pay for informal employment than in more affluent areas, but no informal factories or businesses have emerged to succeed those formal businesses which have closed. Surveys of relatively affluent new towns and commuter areas in France (Cornuel and Duriez 1985, Tievant 1982), in contrast, discover a relatively high amount of informal employment, maintained principally by the more prosperous and professional residents who undertake individual informal employment ventures as much for social as for economic purposes. In Italy, similarly, direct studies of informal employment reveal that it is more extensive in the relatively affluent northern regions than in the more deprived southern regions (Mattera 1980, Mingione 1991). As Mingione and Morlicchio (1993: 424) declare, 'the opportunities of (*sic*) informal work are more numerous, the greater the level of development of the surrounding social and economic context'.

However, despite the prevalence of the finding that informal employment is more extensive in more affluent areas, there are a number of studies which counter this general rule, suggesting that it is not a universal principle (see Williams and Windebank 1997). In a study of five neighbourhoods in Amsterdam, for example, Renooy (1990) finds that although the gentrifying inner-city area of Brouwersgracht conducts a greater proportion of its total work through informal employment than the two areas with high proportions of social security claimants studied, Witte de Withstraat and Deurlosstraat, the two middle-class areas examined, Amerbos and Slotervaart, undertake a smaller proportion

101

of their work using informal employment than the poorer areas. It appears, therefore, that it is localities dominated by middle-income groups that undertake the most informal employment. It could be suggested, therefore, that those with the highest incomes do not need to undertake such work or hire such employees. That said, it is not the case that the magnitude of informal employment can be always and everywhere directly read off from the level of affluence. Instead, there are many other factors that need to be taken into account when considering what shapes the intensity of informal employment in a locality.

For example, and as we have already seen in the previous chapter on ethnicity and immigration (chapter 6), a locality may have a higher level of informal employment than would otherwise be the case when it is the settlement area of a large illegal immigrant population. However, this is not to say that localities and neighbourhoods so populated inevitably have higher levels of informal employment than others. As shown in chapter 6, this is dependent upon a range of additional factors, such as the socio-economic mix of the area, the background of the immigrant population itself, the nature of the local labour market and the way in which the immigrant population is viewed by its neighbours. All of these will influence whether the illegal immigrants take informal jobs, what informal jobs are open to them or whether they adopt alternative strategies to insert themselves into formal employment.

Similarly, chapter 5 has shown that an area may have a higher level of informal employment if local labour market flexibility with regard to women's employment is achieved through informality, than a local labour market in an area where low-level, low-paid, part-time or temporary formal jobs fulfil this function (Cousins 1994). Which structure prevails in a local labour market is dependent not on the level of affluence in the area but on a diverse range of additional conditions. Poverty, in sum, is unable on its own to be a predictor of the level of informal employment in a locality. Although economic, social, institutional and environmental conditions frequently combine in the advanced economies to produce a 'cocktail' which means that middle-income areas conduct greater levels of informal employment than poorer areas, this is not everywhere and always the case.

Examining how poverty influences the character rather than magnitude of informal employment in localities and regions, there is again a general finding that higher-quality informal employment is concentrated in more affluent localities. Put another way, deprived areas tend to undertake a higher proportion of poorly-paid, exploitative, organised, informal employment whilst relatively affluent areas tend to undertake more well-paid autonomous informal employment. Mingione (1991), for example, distinguishes three Italies: the more affluent and industrialised north-west, in which autonomous informal employment is widespread; the north-east and centre, in which although informal employment has developed in order to cut production costs, there is both organised and individual informal employment conducted under reasonable conditions as well as low-paid informal work; and the south, in which informal employment is more exploitative in character than that found in the centre or north of the country and where some of the most exploitative forms of informal employment (e.g. industrial homeworking, illegal manufacturing of forged fashion items) are concentrated.

Similar patterns are identified in Spain so far as the spatial variations in the character of informal employment are concerned. Examining how displaced workers previously employed by large firms were forced to seek jobs in smaller underground businesses or to do informal piece-work in their own homes, Benton (1990) detects a qualitative difference in the experiences of such workers in poorer and richer parts of the country. Workers previously employed in the electronics industry around the richer area of Madrid became skilled worker-entrepreneurs, taking on sub-contracted informal employment indirectly from their previous employers, frequently for high rates of pay. In contrast, those displaced from the shoe industry in the poorer region of Alicante witnessed a general degradation in working conditions and pay. This is further reinforced by studies in the deprived areas of Andalucia, Sabadell and Valencia where informal employment has been found to be predominantly of the organised exploitative kind (Lobo 1990a). Such findings have been echoed in Greece (Hadjimichalis and Vaiou 1989, Leontidou 1993, Mingione 1990).

However, and akin to studies of the volume of informal employment, it is not always the case that exploitative informal employment predominates in deprived areas and autonomous informal employment in relatively affluent areas. Leonard (1994) for example, identifies fairly high levels of autonomous informal employment on a deprived housing estate in West Belfast (see chapter 4) and several studies of deprived rural areas dominated by declining heavy industry and with high unemployment rates again show a good deal of autonomous work (Jessen et al. 1987, Van Geuns et al. 1987, Weber 1989). So, the picture is a complex one. There appear to be other factors besides the level of poverty which determine the extent and character of informal employment in a locality, region or nation.

Beyond deprivation as an explanation for the geography of informal employment

On the basis of the evidence from these direct studies of informal employment in particular localities, it can be suggested that no straightforward universal correlation exists between level of affluence and volume or character of informal employment. Although poorer nations appear to engage in more informal employment than richer countries, it is not generally the poorest populations in the poorest localities and regions of these countries who participate most extensively in this activity. That said, there are exceptions to this general rule. The same can be said of the nature of informal employment. On the whole, although poorer localities are revealed to engage in more informal employment of the organised exploitative kind and affluent localities greater amounts of autonomous informal employment, this is not everywhere the case. Consequently, deprivation alone cannot explain the spatial variations in the character and volume of informal employment. Other factors must come into play which facilitate or constrain informal employment and lead to these more complex configurations of the phenomenon.

In consequence, we here argue, similarly to earlier chapters, that the local character and volume of such employment is the outcome of a 'cocktail' of factors. This cocktail is a mixture of a range of economic, institutional, social and environmental influences which combine in different ways in different places to produce particular local configurations of informal employment. As Mingione (1990: 42) asserts in relation to the spatial variations in Greece, 'Everywhere, the mixes appear to be very complex and fundamentally based on a range of combinations of local opportunities.' The temptation, therefore, is to compensate for the previously overgeneralised and simplistic typologies concerning the geography of informal employment by concluding with such an idiographic approach. We could conclude, similar to Smithies (1984: 89), that 'Black economies are as different from one another as "legitimate" economies.' However, to leave it at this would be to go too far to the other extreme, ignoring some important parallels and commonalities in the extent and character of informal employment between localities and regions. Just as one can discern general tendencies in the spatial divisions of formal labour (e.g. Massey 1984), so too are there patterns to the spatial divisions of informal labour. However, just as in formal labour markets, 'Even "common" dynamics take on different forms in different places, with the result that they are associated with different concrete outcomes' (Peck 1996a: 265), the same applies, as we shall now see, to informal labour markets.

Towards a typology of localities

Here, we present a tentative typology of localities according to the magnitude and character of their informal employment. Four locality types are distinguished according to whether they undertake relatively high or low levels of informal employment and whether such work is mostly autonomous or exploitative in character (see Figure 7.1). Taking each locality type in turn, we first analyse the cocktail of economic, social, institutional and environmental conditions prevalent in each type of area and, second give examples of where such localities are likely to be found in the advanced economies. In so doing, we provide an evaluation of the processes which lead to a particular configuration of informal employment in each type of area and an analysis of the patterns of these spatial divisions of informal labour in the advanced economies.

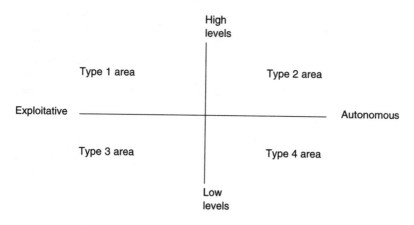

Figure 7.1 Typology of localities relating to the magnitude and character of their informal employment.
Source: Williams and Windebank (1995a: Figure 1).

Type 1 areas: high levels of mostly exploitative informal employment

Type 1 localities are those with high levels of informal employment but mostly of the more exploitative kind. Starting with their economic characteristics, they are inhabited by relatively deprived populations and have an industrial structure marked by a preponderance of small firms whose principal competitive strategy is the reduction of production costs. This economic context both provides a supply of potential informal workers and creates a demand from firms for such a form of employment. They also tend to be areas in which there are high tax and social contribution levels and a tax structure that is biased towards employers' contributions for raising revenue, thus providing yet more incentive for employers to shift towards employing informal workers.

Furthermore, the social conditions in type 1 areas need to be conducive to informality. First, these areas are often characterised by dense social networks that facilitate the easy matching of labour supply to demand, and their social norms and moralities view such employment as an acceptable part of the social and economic fabric. These areas, in addition, are normally characterised by low education levels.

Moreover, these areas have institutional characteristics which foster informality. First, their populations usually have poor access to permanent state benefits and the authorities are lax in their enforcement of rules and regulations in the realms of both social benefits and labour law. Indeed, authorities in type 1 areas passively tolerate the existence of informal employment (e.g. Lobo 1990b, Van Geuns *et al.* 1987) and also frequently actively support informal employment so as to enable firms to compete in international markets and/or help individuals and families raise adequate incomes, which would be difficult if 'regular' regulations were strictly adhered to (Cappechi 1989, Lobo 1990a, Portes and Sassen-Koob 1987, Van Geuns *et al.* 1987, Warren 1994). These type 1 areas also commonly possess certain environmental characteristics. They are often located in deprived rural areas and/or tourist areas, although this is not exclusively the case. They may also exist, for example, in those neighbourhoods of cities in which immigrant populations are concentrated.

Examining those studies of localities and regions which identify this combination of characteristics, the overall conclusion is that type 1 areas are most likely to be found in the poorer regions of southern EU nations, especially in rural areas, or in ethnic enclaves in US cities. Examples are seen in investigations of central, north-eastern and southern Italy (Mingione 1991), Alicante (Benton 1990), Andalucia, Sabadell and Valencia (Lobo 1990a) and Greece (Hadjimichalis and Vaiou 1989, Leontidou 1993, Mingione 1990), as well as in the case studies of ethnic enclaves in US cities such as Miami and New York discussed in chapter 6. All of these locality studies describe areas with high levels of organised exploitative informal employment and show that these areas possess economic, social, institutional and environmental conditions very similar to those analysed above. They also show that the mere existence of one or other of these characteristics on its own is insufficient to produce this type of area. Instead, these locality studies clearly depict that it is the way in which all of these economic, social, institutional and environmental characteristics combine which creates this particular configuration of the informal labour market.

Type 2 areas: high levels of mostly autonomous informal employment

Type 2 localities, which have high levels of informal employment but mostly of the more autonomous kind, are similar to type 1 areas so far as their social characteristics are concerned. They often possess conducive social norms and moralities making such work acceptable, dense social networks and a broad socio-economic mix. However, these social characteristics which facilitate a high level of informal employment combine with very different economic, institutional and environmental conditions in type 2 areas compared with type 1 areas, which explains why their informal employment tends to be autonomous rather than exploitative.

The nature of these other characteristics varies in different kinds of type 2 area. The first kind of type 2 area, usually found in southern EU nations, differs from type 1 areas in terms of the prevailing economic conditions in the sense that they usually have an industrial structure characterised by a preponderance of small firms; these firms are pursuing high value-added production through technological innovation rather than merely engaging in low value-added production and seeking to lower production costs as much as possible. These areas also usually have relatively high levels of affluence, low unemployment rates, a relatively highly skilled labour force that is well qualified and a lax attitude to enforcement of rules and regulations by the state authorities. The outcome is thriving enclaves of small informal enterprises that have adapted to changing market demands for specialised products while sustaining relatively high wages and good working conditions. Examples of such areas are seen in the writing of Mingione (1991) on the affluent north-west of Italy, the work of Cappechi (1989) on Emilia Romagna and the study of Benton (1990) on the electronics industry around Madrid.

The second kind of type 2 area, which is more prevalent in northern EU nations and North America, again possesses the social characteristics discussed above but differs in other respects to the type 2 areas most commonly found in southern EU nations. To see this, we distinguish two contrasting varieties of type 2 area found in northern EU nations and North America. On the one hand, there are type 2 localities characterised by deprived and unemployed populations similar to those in type 1 localities, but in contrast, due to the existence of dense social networks not dependent upon employment, and perhaps alternative activity being available such as in neighbouring rural areas, these populations can overcome their deprivation with relatively autonomous informal employment carried out on an individual basis. Examples are the old mineworkers' colonies in the Netherlands (Van Geuns *et al.* 1987), including Landgraaf in the eastern mining region (Renooy 1990), as well as the ex-steel industry community of Grand Failly in rural France (Legrain 1982).

On the other hand, type 2 areas can also be found in relatively affluent localities in the northern EU nations and North America. These may be gentrifying areas, as Renooy (1990) found in Amsterdam and Sassen (1991) in New York, where there is a diverse socio-economic mix which provides both the demand and supply necessary for high levels of informal employment. Alternatively, they may also be found in new towns or middle-class areas with many incomers where the population is seeking to create denser social networks outside the employment-place and thus engages in autonomous informal employment primarily for social reasons. Examples are to be found in the work of both Komter (1996) in the Netherlands and Cornuel and Duriez (1985) in France. All of these types of area conduct relatively high levels of mostly autonomous informal employment.

Type 3 areas: low levels of mostly exploitative informal employment

Type 3 areas have low levels of informal employment and that which does exist is mostly of the exploitative kind. Commencing with their economic characteristics, these areas tend to be deprived localities with high unemployment levels and a high degree of economic vulnerability so far as the industrial structure is concerned. Furthermore, they are areas usually characterised by little support or sympathy for people (especially the unemployed) working informally, making such work unacceptable and causing people to fear being reported if they engage in informal employment. These areas also tend to possess a poor socio-economic mix and to display below average educational qualifications. As far as the institutional context is concerned, such areas are governed by authorities exercising tough rules and regulations concerning benefit fraud, tax evasion and labour laws, which they enforce strongly. In environmental terms, such areas are also frequently either socially or geographically isolated and have low levels of private ownership of housing.

Analysing those studies of localities and regions which identify this combination of characteristics, the overall finding is that type 3 areas are most commonly found in poor inner-city neighbourhoods and public sector housing estates in northern EU nations. For example, the discussion of Protestant East Belfast by Howe (1988) comes close to describing such a locality, as does the examination by Barthe (1988) of Orly-Choisy, the study by Foudi et al. (1982) of the Lille area, both in France, the investigation by Morris (1995) of Hartlepool and the survey by Jordan et al. (1992) of a deprived estate in south-west England. No type 3 areas have been identified in the US literature so far as is known, perhaps because most studies of deprived localities have been focused upon only those areas with high levels of ethnic minorities and immigrants, as indicated in chapter 6.

In these areas of high unemployment, therefore, the population possesses the free time but lacks the additional resources and opportunities necessary to engage in a wide array of informal employment. Individuals lack the skills, tools and materials to engage in such work and the social networks to hear about informal employment opportunities, fear being reported to the authorities and have less scope for undertaking such work in the areas in which they live. For these people, in consequence, not only are formal employment opportunities few and far between but informal employment also appears to pass them by. The few opportunities for informal employment which do exist, moreover, tend to be in the form of exploitative organised informal employment (e.g. contract cleaning, fruit picking, unlicensed taxi-driving, labouring, kitchen work). This would not perhaps be problematic so far as the geography of social inequality is concerned if these type 3 areas occupied only a minor space in the advanced economies. The serious issue for socio-spatial disparities, however, is that this is the dominant way in which informal employment is configured in the majority of deprived areas of the advanced economies, as shown earlier in the chapter. This has major implications for socio-spatial divisions not only in informal employment but standards of living in general. Put another way, informal employment does not appear to be a prominent vehicle used in deprived localities to cope with poverty, nor does it appear to reduce the spatial inequalities produced by formal employment.

Type 4 areas: low levels of mostly autonomous informal employment

Type 4 areas have low levels of informal employment but where it does exist, it tends to be mostly of the autonomous kind. Economically, these areas tend to be very affluent localities and neighbourhoods with very low unemployment rates. As we have seen above, it is middle-income, or *relatively* affluent, areas which tend to have the highest levels of autonomous informal employment. Type 4 areas, however, reach a level of affluence where the extent of informal employment starts to tail off as formally produced goods and services replace informal provision. Furthermore, the social fabric of type 4 areas is often composed of a very poor socio-economic mix. Instead, there are mostly affluent households with a well-educated population and little social cohesion in terms of dense social networks amongst the population. The nature of the population precludes, therefore, paid informal activities being used for social cohesion purposes. Moreover, these areas are frequently socially or geographically isolated in the sense that there are few low-income areas nearby or there are major social barriers preventing the transfer of informal work across these boundaries. Such areas are also usually embedded in an institutional context comprised of good access to permanent welfare benefits and strict enforcement of state rules and regulations, including those on benefit fraud and tax evasion, which reduce the supply of informal labour to these affluent households.

The little informal employment that does take place, nevertheless, is mostly autonomous in character. This autonomous informal employment is of two varieties. On the one hand, there is activity such as child-care, domestic help and house repair and maintenance carried out, usually by outsiders, for the people who live in these areas. On the other hand, there is informal employment conducted by the residents themselves, often as a direct offshoot of their professional activities. On the whole, nevertheless, very little informal employment takes place. Instead, and in comparison with other areas, most goods and services are acquired through formal employment.

Examining those studies of localities and regions that identify this combination of characteristics, the overall finding is that type 4 areas are most commonly found in affluent localities and neighbourhoods in or around the urban areas of the advanced economies. An example would be a wealthy suburb of an affluent city, such as Laren in North Holland (Renooy 1990). However, it must be said that few studies have considered such affluent areas in their investigations.

Caveats regarding the typology

Having outlined this typology, several comments need to be made. Not only may other factors combine in different ways to those stated above to produce the same locality-type but the absence of just one of these factors, even if only temporarily suppressed, may change the nature and magnitude of informal employment in a locality. Conversely, the lack of one of these factors may well be compensated for by the greater intensity of another. Hence, some care needs to be taken not to use this typology as an absolute and invariable description of the characteristics possessed by localities that conduct high or low levels of mostly autonomous or exploitative informal employment. In addition, this typology must not be assumed to imply that certain locality-types are spatially segregated into different regions of the advanced economies. Indeed, different types of locality may well be only a few kilometres away from each other. For example, one can well imagine the existence in a northern EU member state of a type 2 area (e.g. a semi-gentrified district of a city) just a few kilometres away from both a type 3 area (e.g. a deprived inner-city council estate) and a type 4 area (e.g. a very affluent suburb). Similarly, in US cities, such areas may well exist within a few blocks of each other, such as in the inner-city districts of major urban areas.

Until now, moreover, we have only discussed this typology in static terms. To understand the character and volume of informal employment, however, a dynamic element needs to be built into such a model of the geography of informal employment. Spatial divisions of informal labour, that is, are similar to spatial divisions of formal labour as described by Massey (1984) in that they are the result of a combination of layers of the successive imposition of new rounds of investment and disinvestment, and any change in a factor (e.g. taxes) may have different effects in different areas, since it comes into contact with different pre-existing structures. Furthermore, these local structures do not passively receive whatever is imposed from outside but can, in turn, interact with, and to some extent, moderate or modify these external forces differently. The magnitude and character of informal employment, therefore, is the product of how local economic, social, institutional and environmental conditions, produced by the impacts of previous bouts of restructuring on the locality, combine in multifarious 'cocktails' to produce further unique local outcomes.

Finally, it should be noted that the individual locality studies do not produce comparable material, so the concepts of 'higher' and 'lower' levels of informal employment have to be used with great care. For example, although southern Italy engages in less informal employment compared with other regions of Italy, no existing studies allow us to compare it with deprived localities in other nations in terms of the volume of informal employment to see if one is higher than the other. Moreover, we do not know whether the differences between nations in their suggested levels of informal employment, for example, are due to the richer populations and localities of these nations alone doing greater amounts of such work than their counterparts in other nations, whether they are due to their poorer groups and localities being engaged more heavily in such employment, or whether higher levels of informal employment are present in both rich and poor regions of these nations. It is the poverty of research which thus limits the further development of this typology.

Conclusions

This chapter has revealed that although the indirect macro-economic methods suggest that poorer nations engage in more informal employment than relatively affluent nations, the direct surveys of informal employment show that it is not generally the poorest localities and regions with high concentrations of marginalised people who participate most heavily in this activity. Instead, many of the direct surveys reviewed in this chapter have revealed that informal employment is sometimes greater in more affluent localities. Upon examining the character of informal employment, moreover, poorer areas have been shown to be more likely to engage in relatively exploitative organised informal employment whilst affluent areas often appear to engage in greater proportions of autonomous informal employment. Thus, one might conclude from these findings that contrary to popular prejudice, informal employment is inversely related to the level of poverty in a locality and/or region. The problem, as shown, is that this general finding concerning the spatial divisions of informal employment does not always apply. Important exceptions to this general rule are prevalent. Consequently, poverty alone cannot be used to explain the spatial configurations of informal employment. Instead, other factors come into play that constrain or facilitate the nature and extent of such work in an area.

Here, we have argued that the nature and extent of informal employment in any locality is the outcome of a 'cocktail' of factors, composed of a range of economic, institutional, social and environmental influences, which combine in various ways in different localities to produce particular local outcomes. To explain how these factors combine, a tentative typology has been presented of the characteristics of localities that will undertake high or low levels of mostly autonomous or exploitative informal employment.

However, this typology must not be seen as a concrete theoretical model that can be used to map the magnitude and character of informal employment across the advanced economies. Despite the volume of literature on informal employment, what remains striking is the poverty of in-depth micro-social studies of particular localities as well as cross-national comparative studies of this phenomenon.[2] Having used existing studies to sketch a tentative typology of the spatial variations in the magnitude and character of informal employment, what is now required is in-depth research into localities representative of these 'locality-types' so as to map out in a more systematic and comparative manner than has previously been the case the configurations of informal employment in different countries, regions and localities. Until this is completed, our understanding of the spatial divisions of informal employment must remain only tentative.

8

INFORMAL EMPLOYMENT IN DEVELOPING NATIONS

Introduction

The origins of the study of informal employment in the so-called 'developing' nations lie in the work of Hart in Ghana over twenty-five years ago (Hart 1973), thus mirroring the interest shown in this form of employment in the advanced economies at about the same time (e.g. Gutmann 1978, Henry 1978). Despite equally long histories, there has been relatively little exchange of knowledge between the two camps. Instead, the two literatures remain disconnected, with commentators rarely venturing into each others' territory. This is doubtless because of the vast size of each of these two sets of literature as well as the fact that they emanate from within separate disciplines. It is also because in most topic areas, there is a deep divide between the study of developing and advanced economies. Informal employment is no exception. Here, therefore, we intend not only to set foot in a realm seldom entered by analysts who study the advanced economies but also to seek out whether there are similarities concerning informal employment in developing nations both in the way in which the subject is theorised and the phenomenon on the ground. To do this, we pose the same key questions in relation to the developing economies as have already been addressed with reference to the advanced economies.

Are the developing economies witnessing formalisation, as modernisation theory would have us assume, or is there a process of informalisation taking place? Alternatively, is such an either/or choice too simplistic to capture the diverse experiences of this large portion of the total world population? Moreover, is the 'marginality thesis' applicable to informal employment across the entire developing world? Is such employment merely an exploitative form of peripheral work existing amongst marginalised groups or is there a heterogeneous informal labour market with a hierarchy of its own? If so, does it mitigate or reinforce the socio-spatial divisions prevalent in the formal labour market? In asking these questions concerning the magnitude and character of informal employment, our intention is not necessarily to answer them comprehensively. Instead, it is merely to explore whether many of the theorisations of informal employment arising out of our study of the advanced economies are paralleled in the developing economies. To analyse this, we evaluate first, the volume and growth/decline of informal employment and second, the character of such employment in these areas.

The changing magnitude of informal employment in developing economies

The formalisation thesis asserts that as economies develop and mature, there is a shift of economic activity from the informal to the formal sphere. Indeed, much of the discourse in economic development is so embedded in this theorisation that formalisation is often the 'measuring rod' used to define an economy as either 'modern' or 'backward' (e.g. Rostow 1960). However, there are good reasons for believing that this trajectory of economic development is neither natural nor an inevitability. First, there is the evidence already presented in chapter 3 from the advanced economies. This shows that the culmination of formalisation – full employment and a comprehensive welfare state – can no longer be accepted as the end-state of economic development even in these wealthy nations. Rather, it is one way of organising work and welfare which existed in the middle of the twentieth century in a few nations and which has since been transcended by a new phase of economic development composed of heterogeneous processes in different places. Second, there is the evidence from the developing economies themselves. Employing a similar spectrum of techniques to those used in the advanced economies, ranging from indirect to direct approaches (see chapter 2), such evidence has been plagued by problems, especially with regard to the more indirect methods. Unlike the advanced economies, and as the International Labour Organisation (1996) asserts, it is often the case that national data are not even available to indicate whether employment is decreasing or increasing, never mind the proportion of it which is formal or informal. In developing nations, therefore, the micro-level direct surveys of informal employment are often the only evidence available to assess its level and character, thus reflecting to an even greater extent the situation in the advanced economies.

This evidence, in sum, suggests that the assumption of a progression from informal to formal economic activity is by no means universally applicable. Instead, there are heterogeneous processes in different localities, regions and nations. Take, for example, East and South-East Asia. This is often assumed to be a region of the world which has undergone a widespread formalisation of economic life due to the new international division of labour (NIDL) which has increasingly dispersed physical production functions into this region whilst retaining the control and command functions in a network of global cities in advanced economies (e.g. Dicken 1992, Sassen 1991). During the 1970s and 1980s, the beneficiaries of the NIDL were the middle-income nations such as the Republic of Korea, Hong Kong and Singapore, who have now purportedly joined the ranks of the advanced economies and been replaced by a second wave of middle-income countries including Malaysia, Thailand and Indonesia (Hall 1996). It appears, therefore, that formalisation is a widespread phenomenon in this region of the world. Indeed, the evidence on employment growth seems to support this assertion. Between 1986 and 1993 in East and South-East Asian nations, with the sole exception of Indonesia, employment rose at more than 3 per cent per annum, well in excess of the rate of increase in the labour force (International Labour Organisation 1996: 143).

However, there is evidence that the process of formalisation may not be as clear-cut as is sometimes assumed in discussions of these 'tiger' economies. For example, examining Hong Kong, Singapore, South Korea and Taiwan, Cheng and Gereffi (1994) highlight the way in which informal employment has played a major role in their recent economic development and growth. In Taiwan, for instance, very weak regulation of the small firms sector has enabled the growth of informal employment which has been a central pillar in the country's success at pursuing export-led development. The problem, therefore, is that the growth of formal jobs in this region cannot be taken as evidence of formalisation. To believe this is to erroneously assume that formal jobs are universally a substitute for informal jobs. As Cheng and Gereffi (1994) clearly show in the case both of the Hong Kong and Taiwanese economies, formal jobs have not replaced informal employment. Rather, they have grown in tandem.[1] Whether the remarkable growth rate of formal jobs in East and South-East Asia can thus be taken as an indicator of formalisation is extremely doubtful. Instead, it appears that there are different processes in different nations.[2] For example, whilst in Hong Kong and Taiwan informal employment seems to have expanded alongside formal jobs, in the more highly regulated economies of South Korea and Singapore, such employment appears to be both limited and, if anything, contracting relative to formal employment.

In Latin America and the Caribbean, similarly, there does not appear to be one distinct development pathway that all nations are pursuing. As Table 8.1 shows, there is little if any evidence in this region of a universal process of either formalisation or informalisation. On the one hand, some nations appear to have witnessed formalisation of their economies during the 1990s, albeit at different rates and to varying extents. In the already heavily informalised nation of Honduras, for example, the share of informal work in non-agricultural employment declined rapidly from 54.2 to 51.9 per cent between 1990 and 1994, whilst in the previously more formalised economy of Panama, the share of informal work declined at only a marginal rate from 40.4 to 40.2 per cent. On the other hand, however, there are also nations witnessing a process of informalisation, again at very different rates and from contrasting base levels. Already heavily informalised Paraguay, for instance, further increased its share of informal work in non-agricultural employment from 61.4 to 68.9 per cent between 1990 and 1994 whilst the more formalised nation of Chile saw informality increase at a slower rate from just 49.9 to 51.0 per cent.

Table 8.1 Share of informal sector in non-agricultural employment, Latin America, 1990 and 1994

	1990	*1994*
Argentina	47.5	52.5
Bolivia	56.9	61.3
Brazil	52.0	56.4
Colombia	59.1	61.6
Costa Rica	42.3	46.2
Chile	49.9	51.0
Ecuador	51.6	54.2
Honduras	54.2	51.9
Mexico	55.5	57.0
Panama	40.4	40.2
Paraguay	61.4	68.9
Peru	51.8	56.0
Venezuela	38.8	44.8

Source: International Labour Organisation (1996: Table 5.5).

Indeed, Table 8.1 suggests not only that there is a pattern of heterogeneous development in Latin America but that this heterogeneity is increasing in intensity during the 1990s. Of those Latin American nations which are formalising, the countries which are doing so at the quickest rate during the 1990s are generally those with already relatively high levels of formality (e.g. Panama). Similarly, of those Latin American nations pursuing informalisation, it is again the countries which are already heavily informalised which are moving along this development path at the fastest speed (e.g. Paraguay). Therefore, the implication is that there is an increasing polarisation of development trajectories.

In other regions of the world, despite the evidence being patchier, it is sufficient to conclude that there is neither uniform formalisation nor informalisation. In sub-Saharan Africa, the data that exists on formal jobs shows that out of thirteen nations, five have witnessed negative employment growth and in another three job growth has been significantly below the growth rate of the labour force. Of the remaining five which have seen jobs increase at a faster rate than the available labour force, two (Mauritius and Botswana) have displayed significant growth rates in formal employment. Similar heterogeneous trends are identified in the Middle East and North Africa (International Labour Organisation 1996).

Such evidence of heterogeneous development paths is applicable not only on a cross-national level but also on an intranational level. Numerous studies have revealed significant local and regional variations in development trajectories. In Nigeria, for instance, Anheier (1992) compares the cities of Lagos and Ibadan and concludes that the very different development trajectories of informal

115

employment in these two areas are the result of the contrasting urban and regional economies in which they are embedded. A study of six metropolitan areas in north-east Brazil (Rio, São Paulo, Belo Horizonte, Recife, Salvador, Fortaleza), similarly, finds significant differences in the magnitude, growth and character of informal employment (Schuster, cited in Lautier 1994). In a study of Mexico, in addition, the international labour organisation, PREALC, reveals significant differences between Guadalajara, Monterey and Mexico City in terms of the extent and nature of informal employment, which they put down to the rather limited variable of the contrasting industrial structure of the three areas (see Roberts 1990). In Mexico, meanwhile, Martin (1996) identifies significant differences in the extent to which, and how, informal employment is used in household work strategies in urban and rural areas.

In sum, the evidence available on both formal and informal employment suggests that similar to the advanced economies, there are varying processes in different places in the developing world. Formalisation, therefore, cannot be assumed to be evenly and continuously occurring in a universal manner across all localities, regions and nations.[3] Neither, moreover, can informalisation, especially given the fact that this process is not everywhere and always the direct result of a lack of formality. Rather, heterogeneous development appears to be occurring.

Characterising informal employment in developing nations: is there a segmented informal labour market?

Informal employment in developing nations has often been caricatured as a form of work which is the last resort for peripheral populations, such as the poor, women and migrants, who are obliged to perform it as a means of survival (Fashoyin 1993, Lagos 1995, Lubell 1991, Maldonado 1995). For example, Fashoyin (1993: 90) argues that 'the informal sector has demonstrated that it can serve as employer of last resort' whilst Lubell (1991: 12) concludes that 'informal sector participants usually constitute the vast bulk of the urban working poor so that informal activity is . . . often a last resort for urban survival'.

Here, however, we question the validity of this marginality thesis. Arising out of our findings in the advanced economies, we first examine the view that informal employment is a form of marginal work and second, explore the configuration of the informal labour market in developing nations according to employment status, gender, migration and its geographical distribution. The objective, in so doing, is to question whether the informal labour market in developing nations is really a sphere for the marginalised or whether it is more accurate to view it as a heterogeneous form of work that has a hierarchy of its own which normally reflects and reproduces the socio-spatial inequalities prevalent in formal employment.

116

Informal employment as marginal work: a critical evaluation

The origins of the formal/informal dichotomy in the development studies literature, to repeat, is usually accredited to Hart (1973) who used it as a way of understanding the vast amount of activity taking place which was at odds with conventional wisdom in 'western discourse on economic development' (Hart 1990: 158).[4] For him, informal work was not solely composed of marginal activities and neither was it a remnant from some pre-capitalist period which would disappear as 'development' ensued. Instead, it was composed of a diverse range of activities and had its own dynamic in the sense that such work was seen as a contemporary phenomenon rather than as an anachronism. Subsequently, however, this heterogeneous and dynamic view of informal employment was quickly lost as the concept became not only institutionalised into and operationalised by international organisations such as the International Labour Organisation, PREALC and the World Bank but also simplified by a range of commentators.

For many, the view has been that there is a separate informal 'economy' or 'sector' inhabited by the marginalised who use such low-paid work as a means of survival (Fashoyin 1993, International Labour Organisation 1972, Lagos 1995, Lubell 1991, Maldonado 1995). For example, the International Labour Organisation (1972: 23–6) distinguished the informal 'sector' as possessing the following characteristics: (a) ease of entry; (b) reliance on indigenous resources; (c) family ownership of enterprises; (d) small scale of operation; (e) labour-intensive and adopted technology; (f) skills acquired outside the formal school system; (g) unregulated and competitive markets. The formal sector, by contrast, was seen to possess the opposite characteristics. This 'dual economy' thesis, however, has subsequently been extensively criticised both in the advanced economies, as we saw in chapter 1, and in developing nations. That is, although general differences in wage rates, contractual status, ease of access, protective legislation and security, capitalisation and size of operation are to be found between formal and informal employment, the above characteristics are no longer viewed as exclusive to informal employment and are seen to imply a misleading homogeneity of formal and informal employment (Bromley and Gerry 1979, Connolly 1985, Dasgupta 1992, Lautier 1994, Meagher 1995, Peattie 1980, Portes *et al.* 1986, Rakowski 1994, Sharpe 1988, Tokman 1978). Instead, informal employment is increasingly viewed as a form of economic activity which is part of a larger structure that also includes formal markets (Bromley and Gerry 1979, Frank 1996, Peattie 1980, Richardson 1984, Sanyal 1991, Tokman 1978). In this sense, the definition and conceptualisation of informal employment has taken a similar path in both the advanced and developing economies. Both sets of literature have recognised the heterogeneity of such employment, its interdependent relationship with formal employment and its non-traditional nature (e.g. Castells and Portes 1989, Rakowski 1994).

Indeed, much of the evidence is that informal employment can no longer be seen as marginal work solely for those excluded from formal jobs who use it as a last resort. Although informal wages are on average lower than formal ones (see De Pardo *et al.* 1989, Guisinger and Irfan 1980, Roberts 1989, 1990), there is a significant overlap in the wage rates of formal and informal workers. Examining the distribution of monthly incomes in formal and informal employment in Lima, for instance, Table 8.2 shows that although the mean formal income is higher than the mean informal income (392,379 soles compared with 263,458 soles), informal incomes often match or exceed those of formal incomes. So, despite informal incomes being skewed towards the lower end of income levels to a greater extent than formal incomes, with 63.9 per cent having informal incomes of less than 250,000 soles compared with just 39.1 per cent of formal incomes at that level, nearly one in five informal workers earn above the mean formal income. In consequence, informal employment can no longer be seen as the lowest-paid form of work conducted by marginalised groups as a last resort. Neither can informal employment simply be seen as a way of exploiting cheap labour.

These findings in Lima are not an exception to the rule. Tokman (1986) finds substantial differences in informal income between owners of informal shops, the self-employed, workers in informal workshops and domestic servants in Costa Rica, Colombia and Peru. Indeed, the variation in income is such that, for some categories of informal worker such as shop-owners, earnings are higher than for formal workers. Portes *et al.* (1986) discover much the same situation in Montevideo, Uruguay. In East Asian nations, meanwhile, Cheng and Gereffi (1994) reveal that informal employment is not a phenomenon of unemployment or underdevelopment but instead is highly productive rather than subsistence-orientated, and is dynamically linked to the growth of national

Table 8.2 Monthly income distributions for formal and informal sectors in Lima, 1983

Monthly income (soles)	Formal sector No.	%	Informal sector No.	%
0	17,260	2.08	33,520	7.13
1– 49,999	20,417	2.47	40,433	8.61
50,000–99,999	30,756	3.71	50,914	10.84
100,000–249,999	255,192	30.82	175,487	37.35
250,000 –349,999	174,341	21.05	78,307	16.67
350,000–499,999	134,539	16.25	34,982	7.45
500,000–749,999	104,687	12.64	28,548	6.06
750,000 –1,499,999	80,268	9.69	22,278	4.74
1,500,000–2,999,999	8,525	1.03	4,498	0.96
3,000,000–4,999,999	1,226	0.15	592	0.13
5,000,000 +	911	0.11	287	0.06

Source: Paredes-Cruzatt 1987b; 3, cited in Thomas (1992: Table 4.4).

economies. Therefore, to define informal employment solely as a marginal activity for the poor and excluded would be to do an injustice to the large number of informal workers who earn above the average formal wage and for the considerable number of formal employees who earn less than the average informal wage.

Paralleling the theoretical developments in the advanced economies, therefore, such data seriously calls into question the marginality thesis. Informal labour markets in developing nations appear to be heterogeneous entities with their own hierarchy. So, who earns these high informal incomes and who is involved in the lower-paid informal employment? To answer this, we now explore how the informal labour market in developing nations is configured both socially (e.g. by formal employment status and gender) and spatially.

Informal employment in developing nations: by formal employment status

A common presupposition in studies of informal employment in developing nations is that individuals are either employed formally or informally, not both. This is due to the belief that informal employment is a form of work for those excluded from formal employment. As we have seen in the advanced economies, however, this is not always the case. Individuals frequently work both formally and informally. Is there a similar tendency in developing nations? Do the formally employed also participate in informal employment and if so, does the nature of the work that they undertake differ from the work undertaken by those who are excluded from formal jobs?

The evidence available suggests that some of the formally employed do indeed engage in informal employment. For example, in Latin America, Addison and Demery (1987) estimate that 10–20 per cent of the formally employed also have an informal job, whilst in Yaounde, Berthelier (1993) finds that 8.3 per cent of the formally employed have another job, 85 per cent of which jobs are informal. Yet other studies reveal that although the formally employed often work informally only on a part-time basis due to their formal job commitments, the work which they do is higher-paid than that conducted by those without a formal job, who undertake full-time informal employment at lower wage rates. Simon (1997), for example, shows this to be the case in his study of informal retailing in the city of Kaduna in Nigeria. Here, part-time retailing is practised by formal wage employees such as civil servants to supplement their salaries at higher rates of pay than those earned by the non-employed who have to work long hours to secure sufficient income to get by. As such, he claims that there is a dual informal labour market in this specific context. This indicates, therefore, that a segmented informal labour market exists in which there is a well-paid core informal workforce of people who often also have a job and a peripheral informal workforce of frequently non-employed

119

people who tend to engage in the more exploitative lower-paid forms of such employment.

Equally, however, and just as in the advanced economies, informal employment does not include everybody excluded from formal jobs (see Gilbert 1994, Lautier 1994, Roberts 1990). Instead, informal employment absorbs certain people more easily than others. As Gilbert (1994: 614) puts it, 'if the informal sector has acted as a sponge, there are equally clearly substantial hurdles preventing some unemployed workers from entering it. These barriers are especially high in the more remunerative areas of informal self-employment.' For Gilbert (1994), therefore, it is necessary to distinguish between different types of informal occupation. While it is easy to enter what Lautier (1994) calls 'survival' activities such as begging or boot-blacking, it is much more difficult to get into those kinds of activities which require skills, capital, know-how or contacts.

One question that remains unanswered, however, concerns the contrasting motivations for engaging in informal employment. Until now, the assumption has been that all informal workers are purely economically motivated. There is a need however, in future research, for this aspect of informal employment to be examined in greater depth in the developing nations. What makes both formally employed as well as non-employed workers engage in informal employment and how do their motivations differ? And is this reflected in the types of informal employment that they undertake and, if so, how?

Informal employment in the developing nations: by gender

As in the advanced economies, it is not women who conduct the vast majority of informal employment in developing nations. It is men (Paredes-Cruzatt 1987a, Simon 1997). This is largely because women remain universally responsible for the unpaid informal work (Baud 1993, Lubell 1991, Momsen 1991, Standing 1989), meaning that they not only have little time left over to engage in informal employment compared with men but are often also tied to the home. Where women do engage in informal employment, therefore, it is usually work which fits in with their domestic roles and which mirrors their domestic duties and responsibilities, such as cooking, sewing, domestic service or home-based work (Martin 1996, Miraftab 1996, Momsen 1991). Take, for example, the case of Lima. Women dominate dressmaking industries but hardly exist in furnituremaking, shoemaking and printing industries and non-existent in the metal products industries (see Thomas 1992, Table 4.10). The gender divisions of the informal labour market, therefore, reflect the gender divisions in other forms of work.

This is nowhere more evident than in informal homeworking. Homework is a strategy employed by women to solve the contradiction between gender ideology (i.e. men as breadwinners and women as housewives) and the harsh economic realities that compel women to generate cash income, as shown in

Iran (Ghavamshahidi 1996) and Mexico (Miraftab 1996). Indeed, homeworking is almost a form of work exclusively for women. In Lima for example, just 5.4 per cent of all homeworkers are men (Paredes-Cruzatt 1987a). Consequently, and as Ghavamshahidi (1996: 124) argues,

> homework reinforces a gender ideology that proclaims that women should be at home, taking care of their children and their homes as their primary responsibility . . . Homework as a solution to economic problems in families does not violate gender ideology and serves to maintain the historical, social, and economic relations in households structured by patriarchy.

Indeed, developing country governments often actively reinforce such an ideology. In Taiwan, for instance, the state has simultaneously sponsored 'living-room factories' to promote homework coupled with 'mothers' workshops' to remind women of their primary roles as 'good wives and mothers' (see Cheng and Hsuing 1991). Employers also evoke such an ideology of women's primary responsibility to their family and dependence on men as primary breadwinners in order to justify the low pay and irregularity of homework (Hadjicostandi 1990, Lim 1993).

Men, on the other hand, relatively free of such domestic constraints, tend to engage in forms of informal employment which are often not tied to the vicinity of the home or even community. Take, for example, the rural to urban migration, in many developing nations, of people in search of formal or informal employment. This is frequently a male preserve. As Thomas (1992) asserts, in Africa and most of Asia, the majority of migrants are young males. In parts of Africa, this is because land may be communally controlled and worked in families, resulting in the fact that families cannot sell the land and will receive no compensation if they decide to move to the city. Therefore, the man may move to the city, leaving women and children to work the land.[5]

It is not only the types of informal employment undertaken which are heavily gendered but also the contrasting motivations for conducting such work. Women in developing nations, similar to women in advanced economies, tend to undertake informal employment primarily as a means of meeting family needs without a sense that it could be the beginning of upward occupational mobility, whilst for men it is more frequently seen to be the first step on the rung to formal employment. In addition, women insert their earnings into the family budget whilst men do not, keeping part of their earnings for themselves (Beneria 1992, Miraftab 1996, Safa and Antrobus 1992). These motivations remain gendered in this manner, moreover, even when men work at home. In Mexico, for example, Miraftab (1996) finds that women informal homeworkers often perceive themselves as housewives and their homework as merely helping their husbands and family, whilst male homeworkers view themselves as bread-winners and their homework as no different from work outside the home.

121

Indeed, the time men spend on paid work continues to be perceived as distinct in the same way that their jobs outside the home are seen as distinct.

In sum, women not only engage in informal employment to a lesser extent than men but when they do undertake such employment, the type of work engaged in and their motivations for doing so fit in with their domestic roles and reflect their domestic responsibilities. The consequence is that women earn less than men when they engage in informal employment (Kalinda and Floro 1992, Psacharopoulos and Tzannatos 1992) and the more such work mirrors their domestic roles, the lower is the pay (Hahn 1996).

Informal employment in developing nations: the role of migration

In the advanced economies, particularly North America, a principal focus has been on the role of ethnicity and immigration on the configuration of informal employment. In the developing nations, in contrast, a key focus has been on the issue of migration. For many years, it was assumed that informal employment was a strategy adopted by migrant workers (for example, rural migrants to urban areas) as a stepping-stone to formal employment (e.g. Harris and Todaro 1970). However, a large array of studies refute this suggestion that informal employment is a transitional sphere and dominated by recent migrants (Richardson 1984, Schulz 1995, Sethuraman 1981). Instead, migrant workers are now seen to constitute just one segment of the informal labour force, even if they are overrepresented relative to the rest of the population (Thomas 1992). Nevertheless, informal employment in the developing nations, although not a complete product of the migration of the population from rural to urban areas and the inability of those urban formal economies to absorb this labour supply, is inextricably bound up with this phenomenon.

Moreover, there is some evidence that migrant groups are more likely to be found in the peripheral informal workforce conducting exploitative forms of low-paid informal employment (Boris and Prugl 1996, Schulz 1995). New migrants, that is, often tend to enter the informal labour market at the bottom of the hierarchy engaged in 'survival' work such as boot-blacking, begging, selling cigarettes and garbage picking (Lautier 1994). Other forms of core informal employment, meanwhile such as running micro-enterprises, tend to be taken up by older migrant populations and indigenous groups.

There are, nevertheless, many questions left unanswered. Do migrants progress up the hierarchy of the informal labour market over time? How does this vary amongst different groups and in different places? Do migrants enter informal employment as a matter of choice? What part do social networks and motivations in the form of common ethnic bonds play in their decisions both to migrate and to enter informal employment? What are the constraints on their choices? How does this vary between different groups and between different areas? How do state regulations and laws concerning migration affect their choice?

The geography of informal employment in the developing nations

Is it the case that although poorer nations do appear to engage in more informal employment than relatively affluent nations, as organisations such as the International Labour Organisation (1996) suggest, it is not generally the poorest localities and regions who most heavily participate in this activity? Moreover, is it also the case that poorer areas seem to engage in relatively exploitative low-paid forms of informal employment whilst more affluent areas conduct greater proportions of better-paid more autonomous informal employment? And are there exceptions to this general rule? If this is the case, how can the geography of informal employment be explained?

On an intranational level, the principal way in which the geography of informal employment has been studied in the developing nations is with regard to the relationship between informal employment and settlement size. Do major urban settlements have higher levels of informal employment relative to small cities, or vice versa? Or, put another way, do settlements with higher levels of formal employment have larger or smaller levels of informal employment? Examining the literature, there is some debate on this issue. As Mathur and Moser (1984) review, there are two contrasting views on the relationship between city size and informal employment. Some commentators suggest that informal employment declines as city size rises whilst others suggest the opposite. For the former, as cities grow they become more complex and thus formalised, so the share of informal jobs in total urban employment declines even if the absolute amount of informal employment does not. For the latter, however, larger cities have a sizeable autonomous demand which creates better conditions for informal employment to survive and expand, provides more opportunities for informal employment to utilise effectively those technologies that are made redundant in the formal sphere because of the pace of technological transformation to be found in such cities, and offers higher average incomes as well as a higher level of public amenities, suggesting possibilities for informal employment. Given that there is evidence to support both theses (see Mathur and Moser 1984, Roberts 1989, Simon 1997), it is here suggested that the relationship between settlement size and the relative and absolute magnitude of informal employment is mediated by a whole range of additional economic, social, institutional and environmental conditions. The result is that there is no universal relationship but, rather, diverse relationships in different places and at varying times.

It can be stated, nevertheless, that urban informal employment is growing and rural informal employment is declining, in part due to urban migration (e.g. Huber 1996, Isuani 1985). The intensity with which informal employment is growing in urban areas, however, is by no means uniform. Numerous studies, as discussed above, show significant differences between urban areas both in the magnitude and growth rate of informal employment in different towns and cities (e.g. Anheier 1992, Lautier 1994, Martin 1996, Roberts

1990). What is certain is that such variations cannot be simply explained with reference to a specific variable such as the level of poverty or the degree of migration. Instead, and as argued in previous chapters, there will be a need to take account of a large array of economic, social, institutional and environmental factors, and how they variously combine to produce particular local outcomes, when explaining the geography of informal employment in developing nations.

Besides these spatial variations in the volume of informal employment, there is also the question of how the nature of informal employment is spatially distributed in developing nations. Is it the case, for example, that the lower-paid, more exploitative forms of informal employment tend to be clustered in the poorer areas whilst the better-paid informal work tends to be concentrated in relatively more affluent areas and populations? The indicative evidence from developing economies (Roberts 1989, 1990, Simon 1997) is that poverty alone is a necessary but insufficient condition for explaining the spatial distribution of the nature of informal employment. Again, therefore, there is a need to consider the wider factors that constrain or facilitate not only the volume but also the character of such employment in an area. In the next section, therefore, we begin to unpack the importance of the various factors that shape the character and magnitude of informal employment in the developing nations and help explain the socio-spatial configuration of such work.

Explaining the socio-spatial configuration of informal employment in developing nations

Our review of the direct surveys of informal employment in developing nations suggests that the same explanatory framework based on the notion of a 'cocktail' of conditions can be used to explain the socio-spatial divisions of informal labour in developing nations as in advanced economies. The consequence therefore, is that at any time, there is neither universal formalisation nor informalisation of the developing nations taking place but rather, different processes in different places. Neither is there one unique configuration of the informal labour market which is homogenous across all places. Rather, there are different configurations in different areas. It is a multifarious phenomenon (see Rakowski 1994).

The principal difference between applying this framework in developing and advanced economies, however, is that the key economic, social, institutional and environmental conditions which shape informal employment vary, as will particular local configurations of informal employment. Take, for example, the influence of economic regulators. Developing nations are similar to advanced economies in that economic conditions such as the level of affluence and employment, the industrial structure, sub-contracting features and the nature of the tax and social contributions structure influence the socio-spatial configuration of informal employment. In developing countries, nevertheless, the ways in which they influence informal employment can be significantly different to the advanced economies. For instance, given that most taxes are raised indirectly rather than directly in developing nations, informal employment will by definition be more a product of the failure to pay indirect taxes (e.g. licence fees for kiosks or market stalls) than a result of the evasion of direct taxes. Consequently, moves to increase direct taxes will have less effect on the overall level of informal employment in developing nations than in advanced economies. Any shift of direct taxation from individuals to companies, or vice versa, will also have little influence on the relative balance between organised and autonomous informal employment. Attempts to encourage informal employees to pay indirect taxes as a means of formalising such activity, moreover, have been met with widespread failure. Indeed, many commentators (e.g. Lautier 1994) are now arguing that attempts to formalise informal employment have foundered in almost all cases where this has been attempted (e.g. Burkino Faso, Brazil, Ivory Coast, Colombia) and thus, paid informal activities cannot be formalised for developmental purposes.

Similarly, whilst many institutional regulators have a heavy influence on the magnitude and character of informal employment in the advanced economies, their influence will be generally much less in developing nations. Take, for example, welfare benefit payments. Given the lack of welfare benefit payments in most developing nations, the influence of this regulator on the configuration of informal employment is minimal and thus cannot be used as tool for constraining the non-employed from participating in informal employment. Instead, and due to its absence, informal employment represents both a principal means of survival for many non-employed and is more fully embedded in everyday life amongst such populations than in the advanced economies. In this sense, there are some parallels with those southern European nations with weak welfare states. As a factor, therefore, welfare benefits are of minimal importance in shaping the configuration of informal employment in developing nations, except in the sense that they frequently do not widely exist. Other institutional factors, however, such as state interpretation and enforcement of rules and regulations, are often of greater importance, at least according to some readings, due either to weaknesses in the state's ability to implement them or corrupt state officials taking informal payments for services rendered (De Soto 1989). Indeed, where there is state support for this type of employment, such indulgence will provide fertile ground for a thriving sphere of informal employment.

So far as social regulators of informal employment are concerned, although the list of factors is similar to those in advanced economies, their level of importance and the ways in which they configure such employment can differ significantly. Although social norms and moralities in developing nations influence the character of informal employment for example, they are often a very different set of norms and moralities to those that dominate in the advanced economies. The widespread practice of purdah amongst the Muslim population, for instance, has a profound effect on women's participation in informal employment in many regions, confining them to home-based informal employment if they engage in such work at all. Examining informal retailing, the result is that in localities, regions and nations where purdah is prevalent, this is a male preserve, as displayed in the Nigerian cities of Kaduna and Kano (Simon 1997, Mabogunje and Filani 1981) and the Bangladeshi city of Dhaka (Nurulk Amin 1987), whilst in Lagos, Transkei and Guadalajara, where purdah is not prevalent, it is the preserve of women (Nattrass 1987, Roberts 1989, Sada and McNulty 1981). In consequence, although social norms and traditions influence informal employment, the constitution of these norms and traditions is variable and so too are their effects. Social networks, moreover, can be an important determinant of whether opportunities are available for new migrants to an area, as shown by Hart's (1973) early work on the Frafra in Ghana where co-operation and exchange amongst new migrants was essential to their survival in both town and country. Roberts (1990) argues much the same in Latin America.

To conclude this section, we do not intend to develop a typology of how the various regulators of informal employment in the developing nations combine to produce particular configurations of such work in different areas, as was undertaken in relation to the advanced economies. Given that informal employment constitutes a large segment of the economies of many developing nations and these nations themselves constitute the vast majority of the global economy, this would be the subject of a book in itself. Instead, we wish to conclude this section simply by cautioning against placing too much emphasis on one factor or set of factors over others when seeking to explain the volume and/or nature of informal employment in developing countries. Consequently, rather than argue that it is either principally economic circumstances which are leading to the growth of informal employment (e.g. International Labour Organisation 1996), primarily institutional-focused factors such as specific government regulations (e.g. De Soto 1989), chiefly cultural traditions and norms (e.g. Simon 1997) or geographical factors such as settlement size (e.g. Richardson 1984), we here argue that there is a need to examine all of these factors as well as how they combine together to produce specific local outcomes. The point to reiterate is that it is not simply the presence of a factor that necessarily and inevitably leads to a particular outcome. Rather, it depends on how this factor combines with others present in a locality, region or nation. Indeed, given the size and importance of informal employment in developing nations, as well as the diverse experiences of such countries, there is a strong case to be made that examining informal employment in developing nations is akin to theorising the changing nature of employment in the advanced economies. It is a key subject that is central to understanding economic restructuring but one in which the diverse experiences of different nations have become subsumed by the search for universals.

Conclusions

This chapter has revealed that the theorisations of informal employment in the advanced economies outlined in chapters 2–7 of this book are not only applicable to the developing economies but also appear to provide some fruitful avenues for future research. First, this chapter has shown that it is far too simple to assert that there is either a formalisation or informalisation across all localities, regions and nations in the developing world. Rather, we have asserted that the process again appears to be one of heterogeneous development in which at any time, some areas are formalising their economies whilst others are becoming more informalised. Second, there also appear to be similarities between the configuration of informal employment in the developing nations compared with the advanced economies. That is, just as in the advanced economies one cannot simply view informal employment as a refuge for marginalised populations (the marginality thesis), the same applies in developing nations. Instead, there appears to be a heterogeneous informal labour market with a hierarchy of its own. This hierarchy, moreover, seems to reproduce, rather than mitigate, the socio-spatial inequalities prevalent in the formal labour market.

For the study of informal employment in the developing nations to move forwards therefore, several questions need to be addressed in a more systematic and widespread manner. To what extent is there a formalisation or informalisation of different localities, regions and nations in the developing nations? How is informal employment related to formal jobs? When is it a substitute and when is it a complementary form of employment? How does the nature, type and pay of informal jobs compare with formal employment in different places? What are the motivations for people engaging in informal employment? Are they entirely motivated by the quest to survive or are there other motivations that are also important amongst some social groups? How does this vary across space? What are the key determinants that shape the level and nature of informal employment in developing nations? How do they combine in different ways in different areas to produce specific local outcomes? What implications does this have for policy towards informal employment? Given that informal employment is possibly at least as significant as formal employment for the majority of the population in these nations and that these countries constitute the vast majority of the world, such questions are of crucial importance. They are essential not only for understanding a major segment of the global economy but also for identifying what can be done to improve the well-being of the majority of the citizens of the world.

Part III

WHAT IS TO BE DONE ABOUT INFORMAL EMPLOYMENT?

Evaluating the policy options

9

REGULATING INFORMAL EMPLOYMENT

Introduction

This chapter outlines and then critically evaluates the approach that dominates policy towards informal employment in the advanced economies, namely the eradication of informal employment through more stringent regulations. Superficially, it may seem to many readers that this is the most appropriate policy response. After all, and as shown in Part II, informal employment generally reinforces rather than reduces the socio-spatial divisions produced by formal employment. However, and as we shall see, such an approach is not as unproblematic as it may initially appear. To show this, we commence by describing in detail the main arguments put forward within this approach concerning informal employment itself, and the ways in which these are grounded in a particular vision of the future of work and welfare. Subsequently, we critically evaluate these arguments from the point of view of both their practicability and desirability.

The regulatory approach to informal employment

Adopted in the main by a wide range of analysts who are politically left-of-centre, this approach is best characterised as a spectrum of perspectives each with its own reasons for wanting to eliminate informal employment through more stringent enforcement of the regulations which govern labour, taxation and/or social benefits. At one extreme of this spectrum, one finds those who wish to banish informal employment primarily because the fraudulent behaviour of informal employees and employers deprives the state of revenue that could be used for social cohesion purposes. At the other extreme, one finds those who seek to eliminate informal employment primarily because they believe that it necessarily entails exploitation. In this section, therefore, we discuss in more detail the arguments to be found at these two extremes of the 'regulation' spectrum, bearing in mind all the while that although it is useful to categorise them in this way in order to understand the varying emphases of individual commentators, they are not mutually exclusive. However, before embarking on this discussion of difference, it should be noted that common to all those grouped within this approach is the belief that informal employment should not and does not have to be a survival mechanism for the poor and unemployed because it is possible to return to or create a world of full employment and/or a universal welfare 'safety net'. This being the case, informal employment has no positive role to play in the future of work and welfare.

The 'informal employment as fraudulent' perspective

This perspective, adopted by organisations such as the European Commission (1995a and b), the OECD (1994) and the International Labour Organisation (1996) as well as by many national governments (see, for example, MacDonald 1994, Rowlingson *et al.* 1997, Wenig 1990), stresses the fraudulent nature of informal employment which it views as serious because it has two unacceptable effects. First, informal employment is said to represent unfair competition for formal activity and thus has a deleterious effect on formal employment. Second, informal employment is seen as undermining the welfare state both by depriving it of tax and national insurance income and defrauding it when people are claiming and working at the same time. Consequently, the only significance of informal employment is that it disrupts the smooth and efficient running of formal economic and welfare systems.

In this view, therefore, informal employment is a 'villain' and has no place in an advanced economy. Indeed, in recent years, this view has not only become stronger but the issue of informal employment itself has become much more prominent on public policy agendas. Confronted by empty public treasuries, rising unemployment, recession, austerity policies and increasing demands for social protection, many states have reacted by raising tax rates and reducing social security expenditure so as to diminish public deficits. However, the perception has arisen that confronted by such measures, people are increasingly turning to informal employment, thus thwarting the efforts of government to resolve their fiscal and social crises. This being so, informal employment is a sickness and the cure is strict enforcement of the regulations surrounding labour, taxation and social security (Feige 1979, 1990, Gutmann 1978). In addition, some argue that tougher regulations need to be coupled with other policies such as a reduction in marginal tax rates through base-broadening so as to encourage compliance (Feige 1979, 1990, Mingione and Magatti 1995) or a more comprehensible and simplified tax and social security system so that people realise when they are defrauding the state (e.g. Rowlingson *et al.* 1997).

Although some are of the opinion that tax and social security fraud should receive equal emphasis since they are twin evils (International Labour Organisation 1996, European Commission 1996a and b, OECD 1994), such views are in the minority since the vast majority of nation-states stress the importance of social security fraud, despite tax fraud being more of a problem. In the US in 1986, for example, only $235 million was fraudulently claimed through unemployment benefit compared with $70 billion not being reported to the tax authorities (Roth *et al.* 1992). In the UK, meanwhile, in the tax year ending April 1995, despite £6 billion being recovered in unpaid taxes compared with just £64.56 million recovered from unemployment benefit fraud (1.1 per cent of the total recovered in unpaid taxes), there were 4,247 prosecutions for benefit fraud compared with just 357 prosecutions for unpaid taxes (Cook 1997, Rowlingson *et al.* 1997). Indeed, the UK government has even resorted to a public advertisement campaign to encourage people to inform the social security authorities of suspected benefit fraudsters, especially those who may be working whilst claiming unemployment benefit.

According to MacDonald (1994: 508–9), the reason for this paradox amongst many national governments is to be found in the prevailing ideology of work: 'Such ideology holds that the experience of unemployment must remain unpleasant (to deter voluntary unemployment and to energise the search for work) and be seen to be unpleasant (to reassure those in jobs that unemployment must be avoided)', as is epitomised by the workfare system operated in the US and increasingly in the UK (see Peck 1996b). Indeed, this focus upon eliminating social security fraud rather than tax fraud reflects the widely-held belief that social security fraud is in some way more 'criminal' than tax fraud. In major part, this reflects the semi-legitimate status of tax avoidance and the perception that tax-payers are merely trying to retain 'their' money whilst benefit fraudsters are taking 'other people's' money (see Cook 1989, 1997).

The 'informal employment as exploitative' perspective

Within this perspective, informal employment is criticised alongside all forms of exploitative work which unscrupulous employers oblige a weak and unprotected workforce to undertake. It is within the context of regulating all exploitative employment out of existence that it is proposed to eradicate informal employment (Castells and Portes 1989, Portes 1994). Furthermore, the vast majority of those who adopt this point of view consider informal employment to be a new form of advanced capitalist exploitation (Amin 1996, Castells and Portes 1989, Frank 1996, Ybarra 1989). In other words, such work is seen as a 'revenge of the market' against state regulation and working-class power in a period of heightened international competition.[1] A strong correlation is drawn between informality and the impoverishment of workers since firms are believed to turn informal so as to avoid the costs associated with protective labour legislation. Informality, therefore, is a process designed to reverse the costly process of unionisation and disenfranchise a large section of the working class (often with the acquiescence of the state) in the interest of renewed economic growth (Castells and Portes 1989, Roberts 1991). Since informal employment is an instrument wielded by different participants in the class struggle and the outcome is to alter class structure and privilege, informal labourers are seen to share characteristics subsumed under the heading of 'downgraded labour': they receive few benefits, earn low wages and have poor working conditions.

In this view, therefore, informal employment should be encouraged to move within the bounds of the law and to comply with labour protection rules so as to help eliminate exploitative labour market conditions. As Portes (1994: 433) puts it,

> the way to promote sustained economic development and to avoid the chaos of the uncontrolled market is to implement detailed regulations and have them enforced by a competent bureaucracy immune to profit-taking. This course would lead to the absolute hegemony of the formal sector and hence the reduction of illegal and informal activities to a minimum.

However, as many of these analysts recognise, extensive regulation is itself problematic since it tends to encourage further informal employment. As Portes (1994: 433) asserts, 'state efforts to obliterate it [informal employment] through the expansion of rules and controls can exacerbate the very conditions that give rise to these activities'. Put another way, 'order creates disorder. The formal economy creates it own informality' (Lomnitz 1988: 54). Consequently, 'while . . . the proper realm of long term planning and accumulation is the formal sector, efforts to extend its scope to the entire economy often end up producing the opposite result, namely the expansion of informal employment' (Portes 1994: 433). However, for adherents of this perspective, this is no reason to give up on increasing the regulatory apparatus of the state. As Portes (1994: 433) stresses, 'state regulation creates opportunities for informal activities, but does not give rise *ipso facto* to them'. Instead, for these analysts, it is when extensive regulation is pursued without providing sufficient formal work and welfare provision for these displaced informal workers that informal employment continues to expand. As Lomnitz (1988: 54) argues,

> the degree of formality and the inability of the formal system to satisfy societal needs give rise to informal solutions. If the formal system is able to produce and distribute the goods and services required by all members of society, informal solutions would be less needed and thus less pervasive.

For this approach, in consequence, tougher regulations regarding informal employment need to be coupled with the development of formal work and welfare provision. Without the latter, usually taken to mean the pursuit of full employment and a comprehensive welfare state, informal employment will continue to flourish. Based on the assumption that civil society is no longer capable of meeting the needs of those social groups and areas most in need of welfare provision (Mingione and Magatti 1995, Portes 1994), the message of this approach is that there is little option open to society other than to couple the eradication of informal employment with the pursuit of full employment such as through active labour market policies (Mingione 1994) and failing this, a comprehensive and universal formal welfare 'safety net' which will capture those who remain excluded from formal employment.

In sum, whether the regulation of informal employment is sought mainly due to its 'fraudulent' or 'exploitative' nature, the goal of this approach is to replace it with full employment and, failing this, a formal welfare 'safety net' to cushion the circumstances of those who find themselves economically excluded. We now evaluate critically the practicality and desirability of this perspective in terms of its vision of both the future of work and welfare as well as its policy prescription for informal employment.

Critical evaluation of the regulatory approach to informal employment

As stated in the introduction to this chapter, the regulation of informal employment may superficially seem the most appropriate approach to this form of work which by its very nature 'breaks the rules'. Furthermore, and as we have shown in Part II, informal employment generally reinforces, rather than reduces, the socio-spatial inequalities characteristic of formal employment. However, there are several major problems with pursuing this approach. These relate first to the possibility of creating the 'full employment/comprehensive welfare state' scenario on which this approach is based and second, to the possibility of eradicating informal employment given the fact that this activity is deeply embedded in everyday economic and social life in many areas. Furthermore, the question is raised of whether such a regulatory policy approach, if pursued, would lead to a widening of the socio-spatial inequalities produced by employment, leaving many populations bereft of the means to meet their needs and wants. We thus commence with a critical evaluation of its standpoint on the future of work.

Critical evaluation of its vision of the future of work

As discussed above, this approach believes the future of work to lie in a return to the 'golden age' of full employment. Here, therefore, we evaluate critically whether this is achievable in practice. To do this, it is first necessary to assess the extent to which the advanced economies are moving towards this end. Examining the trajectory of economic development in the advanced economies, there is in fact little to suggest that this is the case. Quite the opposite. As Pahl (1984) intimated well over a decade ago, 'full employment' is just one way of organising work that existed at most for just thirty years or so following World War II in a few advanced economies. Even then, however, it was only full employment for men, not women, and work was never provided which met the aspirations of workers. For example, even in the middle of the supposedly 'golden age' of full employment in 1965, just 65.2 per cent of the EU's population of working age had a job. By 1992, this figure had further fallen to 60.8 per cent (European Commission 1993). Indeed, by 1994, nearly half (44.7 per cent) of Europe's population aged over 15 (33.3 per cent of men and 55.2 per cent of women) were not in jobs (Eurostat 1996: 8). Examining the cross-national variations, moreover, even if the richer northern nations of the EU were able to achieve full employment, where at best just under a third of the adult population are without employment, it is doubtful whether this would be possible in the poorer southern EU nations where around half the adult population do not have a job (Eurostat 1996: 8–9).

Given this shift away from full employment, the only conclusion which can be reached is that it is unrealistic to expect any of the advanced economies to achieve a state of full employment as it was understood three decades ago. Indeed, in order to achieve this, either there would have to be a return to out-dated gender divisions of labour (with men in employment and women at home) or the provision of jobs for all men and women desirous of employment which has never so far been achieved in the advanced economies.

This pursuit of full employment is even more unrealistic when one realises that the figures quoted above on employment and non-employment levels mask the extent to which underemployment is also rising as permanent full-time jobs are replaced by temporary and part-time ones (Nicaise 1996, Thomas and Smith 1995, Townsend 1997). In 1993, for example, almost one in three employees in the EU worked part-time and the 'life-span' of many jobs (even if defined as 'permanent') is rapidly declining. A recent report, for example, finds that the average 'life' of a job in the EU is 4–5 years, which is comparable to the US (European Commission 1996a).[2] Consequently, there is little if any evidence to suggest that 'full employment' or even fuller employment will be realised in the near future or even beyond. As the European Commission (1996b: 28) concludes, 'it is hardly likely that we will return to the full employment of the 1960s'. Reinforcing this, the ageing of the population of the advanced economies means that even if the proportion of people of working age not in employment were to decline, the overall dependency ratio (the proportion of non-workers to workers) will continue to rise (Nicaise 1996). The evidence is overwhelming, therefore, that despite the long-wave formalisation of the advanced economies, the dream of a stable full-employment society guaranteeing incomes and social participation to the vast majority of the population seems to be receding ever further from our grasp. The problem, as Mingione (1994: 4) asserts, is that 'It is no longer possible to rely on decisive extensions of the stable employment base providing guaranteed adequate incomes or on sufficiently high rates of economic growth to compensate for social exclusion always and to the extent necessary'.

Instead, the trend is that employment is becoming ever more concentrated amongst a smaller proportion of households, with jobs being filled by individuals from households where other occupants are already in paid work. Examining the distribution of employment across prime age (20–59) households for thirteen OECD countries between 1983 and 1994, Table 9.1 reveals a decline of households with a mix of non-employed and employed adults and an increasing polarisation of households into either entirely jobless households (the 'no-longer-working-class' households) or households in which everyone is in employment (the 'new workhouses'). Indeed, by 1994, about one in seven of all households in OECD nations were jobless and the jobs available have been disproportionately taken by households where a working adult was already present.

Table 9.1 ranks each nation according to which had the lowest proportion of households with a mixed employment status in 1994, the aim being to indicate which nations have undergone the most far-reaching polarisation of employment opportunities. This reveals that the neo-liberal Anglo-Saxon nations of the UK and US have the widest polarisation of households into either jobless households or the new workhouses. However, all nations (except Canada) have seen a decline in households with a mixture of employed and unemployed

Table 9.1 The polarisation of employment between households, OECD nations, 1983–94

	Jobless households			Mixed employment status households			Households where all in employment		
	1983	1990	1994	1983	1990	1994	1983	1990	1994
UK	16.0	14.3	18.9	30.1	22.0	18.6	53.9	63.7	62.1
US	13.1	10.0	11.5[4]	32.3	24.9	24.9	54.6	65.1	63.6
Germany	15.0[1]	12.8	15.5	32.5	27.7	25.6	52.5	59.5	58.9
Netherlands	20.6[2]	17.2	17.2	39.1	31.9	27.0	40.3	50.9	55.7
France	12.5	14.4	16.5	30.6	28.3	27.9	56.9	57.4	55.7
Belgium	16.4	18.0	19.6	41.8	33.7	28.8	41.8	48.3	51.6
Australia	11.9[3]	14.5[5]	–	32.6	28.3	–	55.8	57.2	–
Portugal	12.7[3]	10.8	11.0	38.3	32.9	32.6	49.0	56.4	56.4
Canada	15.2	12.5	15.1	35.7	37.0	35.9	49.1	50.6	49.0
Ireland	17.2	20.0	22.3	47.3	40.8	36.9	35.5	39.3	40.9
Greece	16.0	16.9	17.6	46.3	40.1	38.9	37.7	43.0	43.5
Luxembourg	10.9	9.3	10.5	47.3	42.1	39.0	41.8	48.7	50.5
Italy	13.2	14.3	17.2	47.4	43.1	42.8	39.4	42.6	40.0
Spain	19.4[3]	15.2	10.8	54.5	51.6	48.1	26.2	33.2	31.8

Source: Gregg and Wadsworth (1996: Table 1).
Notes
1   Data for 1984.
2   Data for 1985.
3   Data for 1986.
4   Data for 1993.
5   Data for 1991.

adults. Of the twelve nations witnessing this decline, eight have experienced rising proportions of workless households and eleven rising proportions of households in which all adults are employed (France being the exception). So even in those nations in which there was not a rise in workless households, the bulk of extra jobs resulted in fewer mixed households rather than fewer workless households.[3]

If the advanced economies are not returning to the 'golden age' of full employment and jobs are becoming concentrated into fewer households, then the next question is whether the formal welfare 'safety net' is increasingly being put in place to protect those households and workers excluded from employment.

Critical evaluation of its vision of the future of welfare

Given the negative attitude towards welfare provision in the neo-liberal nations such as the UK and the US, which Table 9.1 shows also have the deepest levels of social polarisation, we focus attention in this section on how welfare provision is changing in the social democratic-orientated nations of the EU. The logic of this focus is that if universal and comprehensive formal welfare provision is in decline even in those nations which proclaim a wish to protect it, then there is little prospect of its protection and/or extension elsewhere.

Any review of the direction of formal welfare provision in these EU nations gives little cause for optimism. The aim of the 1986 Single European Act, and the Single European Market (SEM) in particular, was to revitalise tired European economies, make industry more productive and promote faster European growth. Recognising that opening up vastly differing economies more fully to one another might lead to a levelling down of social protection in the resulting battle for lower production costs, especially in those Member States not established as high-technology, high-productivity economies (Bennington *et al.* 1992, Rehfeldt 1992), a social dimension to the Single European Act was introduced in the form of the Social Charter. At the outset of discussions of a Social Charter in the EU, however, the perfunctory debate about citizens' rights was quickly transformed into a discussion of workers' rights (Culpitt 1992, Meehan 1993), reinforcing a trend which already existed in many Member States towards a 'bifurcated welfare model' (Abrahamson 1992). This offers some basic protection for workers but little if any to the more marginalised populations in that a dual welfare system is fostered whereby company-based or employment-related welfare schemes take care of those in employment but neglect or exclude marginal and less privileged groups. Thus, and as Bennington *et al.* (1992) suggest, many have been concerned about the evolution of a corporatist model of welfare in the EU in which social rights are attached primarily to employment rather than to citizenship. The argument therefore, is that the Social Charter will exacerbate inequalities between those with and those without employment.

Even in its attempts to introduce workers' rights, however, there has only been partial success. At the Maastricht Summit in December 1991, an attempt was made to forge a common EU social policy by including the tenets of the Social Charter in the Maastricht Treaty. Britain, however, opposed this, deciding that even a treaty of workers' rights was too costly a policy. The then Eleven, none the less, agreed to go ahead with a common social programme to include a maximum working week of 48 hours, and laws on part-time workers' rights and worker consultation. Nevertheless, they could not ignore Britain's presence, hovering as it did on the peripheries of social Europe, offering a cheap labour location as well as privileged access to the SEM for businesses (Cochrane and Clark 1993, Oakley and Williams 1994). Although Britain's resistance within the EU to a rigorous common social policy has now relaxed with a change of government which has sought to sign the Social Chapter, the competition on wage costs from countries outside the EU, such as those in South-East Asia and Central and Eastern Europe, is still putting great pressure on the EU nations to keep down their own costs of social protection. This could well mean that those countries with the highest levels of social protection will eventually be obliged to lower their standards whilst those countries who had envisaged the EU helping them to upgrade their levels of provision could see their hopes dashed. Hence, despite the extensive deliberations concerning social Europe, any enlargement (or even maintenance) of formal welfare provision will be very difficult in the near future or even beyond (Windebank and Williams 1995).

Indeed, the evidence is that such constraints are resulting in cutbacks in formal welfare provision. The Report on Social Protection for the European Commission, for instance, reveals that growth in expenditure on social protection benefits per capita fell or remained static in six out of fifteen Member States during the 1990s compared with the late 1980s (European Commission 1995a). Moreover, and as Reissert (1994) reveals, the notion that social insurance is universally available is far from accurate, even in these countries striving for such a form of welfare provision. Examining the European Labour Force Survey, he finds that in 1990 the benefit recipient quotas (i.e. the number of benefit recipients as a proportion of total unemployed) ranged from less than 20 per cent in Greece, Portugal, Luxembourg and Italy to more than 80 per cent in Denmark and Belgium. However, nowhere has this figure reached 100 per cent. Unemployment compensation benefits, therefore, are by no means universal, mostly due to the unavailability of permanent social benefits.

Leaving aside any desire for citizen's rights, therefore, even a policy of workers' rights is not working. This is hardly surprising. Even Keynes and Beveridge, the founders of the welfare state, recognised that the foundations of social welfare lie in the labour market. For them, full employment, not welfare states, was the key to economic well-being. Full employment meant low demand for social transfers and a large tax base to finance social programmes for the aged, sick and the minority of persons without jobs. Such welfare states were possible only so long as most people found their 'welfare' in the market most of the time (Myles 1996). The demise of full employment thus has profound implications for the future of the welfare state.

Given that neither full employment nor the associated regime of comprehensive welfare provision are likely to be in place in the near future or even beyond, even in the social democratic advanced economies explicitly seeking this arrangement, the immediate question which arises is over the future for work and welfare in the advanced economies. In a world in which such a model is assumed to be still possible, or perhaps in a world where no political party wishes to be the one to admit that it cannot realise such a future for the electorate, then it is obvious that informal employment should be eradicated as part and parcel of the attempt to achieve this state of affairs. However, if one accepts, as the above evidence suggests, that this ideal is unlikely to be achieved, then the question arises of whether it is appropriate to seek the eradication of informal employment without putting in place any alternatives.

Critical evaluation of its approach to informal employment

Leaving aside the problems that would arise from eradicating informal employment posed above, this section asks whether it is in fact possible and/or desirable to eliminate informal employment from social and economic life, whatever the economic and social context. In other words, we discuss whether there are any features of informal employment, independent of the 'work/welfare' scenario in which it is embedded, which militate against its eradication. First, therefore, we examine the extent to which it is practical to try to eradicate informal employment and second, evaluate critically the desirability of such an aim in terms of the implications for socio-spatial inequalities.

Is it practical to try to eradicate informal employment?

A major problem with this regulatory approach is whether it could ever be successful in eradicating informal employment through tougher regulations. On the one hand, this is because there are 'resistance cultures' to such a policy approach in parts of the advanced economies. Many local and regional state authorities do not wish to abolish informal employment and it will be difficult to persuade them to do otherwise in the context of an increasingly competitive international economic system. On the other hand, even for those authorities that do wish to eliminate such work, there are some inherent problems involved in attempting to do so. First, there are 'practical barriers' to achieving such an end since this is a form of work that is so deeply entrenched in everyday social life and economic production. Second, one has to take into account the 'unintended impacts' of tougher rules and regulations.

The first barrier, therefore, is that of 'resistance cultures'. It will be difficult to persuade many local, regional and even state authorities to eradicate informal employment since it is currently used as a strategy both for promoting economic competitiveness and providing a means of getting by for those who find themselves excluded from the formal labour market and welfare provision. This is especially the case in southern EU nations. During the 1970s and 1980s, for example, manufacturing industry in many southern EU nations turned towards organised informal employment as a competitive strategy in order to free businesses from restrictive labour laws and corporatist work practices (Benton 1990, Cappechi 1989, Mingione 1991). Moreover, state authorities in numerous localities, regions and nations became heavily implicated in these informal employment practices (Cappechi 1989, Van Geuns *et al.* 1987, Portes and Sassen-Koob 1987, Warren 1994). Indeed, this has sometimes been an active economic development strategy. As Vinay (1987) reports, in Italy both the Communist Party in the regional government of the central regions and the Christian Democrats in the north-east have collaborated in the informal practices of small and medium-sized enterprises in their regions through their influence on industrial relations. Such situations have also been reported in Spain (Benton 1990, Lobo 1990a, Recio 1988).

More usually, however, it has simply been a case of lax enforcement rather than active promotion that has been the strategy adopted by state authorities relying on such work. In Greece, for example, Mingione (1990) reports how industrial homeworking legislation is not enforced. In 1957, a bill was approved for the Greek clothing industry, compelling employers to pay social insurance contributions for homeworkers. This was never implemented. According to another bill, approved in 1986, all homeworkers must be considered as wage-workers. This also was not enforced due to employer opposition as well as opposition from the self-employed. Lobo (1990a and b) reveals a parallel lax attitude in Spain and Portugal. In Portugal, for example, Lobo (1990b) argues that interviews with unionists and public administrators reveal that the dominant government attitude is one of 'tolerance'. This is for three reasons: first, because it is seen to help the Portuguese economy compete in a world market where regular conditions would make them uncompetitive; second, because the welfare state is weak and this sector gives many households a level of income not possible if they relied on solely formal employment; and third because inspection of employment-places is inefficient and they cannot control such work.

How to overcome such active promotion and/or lax enforcement of employment regulations is a difficult question. Most northern EU nations desire stronger regulation, especially since the opening of borders following the Single European Market means that such work is now directly in competition with that of their own businesses. Ultimately, this is a delicate matter of national sovereignty, because in the majority of cases we are talking about the enforcement of national legislation and the protagonists in the 'fraud' all want to or must continue with it for their survival. The EU, moreover, cannot police laws when no cases come before the European Court and if it introduces

tougher penalties or tries to toughen up on the policing of informal employment in order to reduce unfair competition, it may find that there is not
the 'political will' in the southern EU nations to comply with and implement
such policies. The issue, therefore, is essentially one of whether the political
will, especially in these southern EU nations, exists to eradicate informal
employment.

It is not simply the lack of 'political will' in the form of these resistance cultures, however, which remains an obstacle to the eradication of informal
employment. As discussed above, there are also practical barriers that must be
taken into account. We have stressed throughout this book how informal
employment is deeply embedded in everyday social life, and is often undertaken
as much for social as for economic reasons. As Komter (1996) reveals in the
Netherlands, to take just one example, such exchange is more frequently undertaken by the relatively affluent as a means of maintaining and solidifying their
social networks. Quite how tougher rules and regulations could put a stop to
such 'social' activity is not immediately apparent since in its most autonomous
manifestation, informal employment is merely a monetarised form of community exchange which could equally well be undertaken within a relationship
of barter or gift.

Furthermore, there is much evidence that any attempts by the state to regulate informal employment are unacceptable to the populace. In Quebec, for
example, Fortin *et al.* (1996) find that 14.7 per cent of the population believe
that informal employment is morally acceptable and 42.7 per cent perceive it as
neither moral nor immoral, whilst just 31.0 per cent see it as immoral. With
less than a third of the population being opposed to such a form of work, it is
thus questionable whether attempts to eradicate it could ever be successful.

To a large extent, this attitude towards informal employment means that
tougher rules and regulations will be problematic to introduce in practice.
Much of the theoretical work, for example, shows that there is a trade-off in
policy-making between the probability of detection and the penalty imposed.
Given the high cost of increasing the probability of detection, the tendency is
to increase the cost of the penalty. However, such policies may not be politically viable given public attitudes. Punishments cannot be set in isolation from
other crimes and there is likely to be strong public reaction if the penalty for
working informally is thought be unfair in relation to others. Consequently,
there may be limits to the range and degree of the penalties authorities can
implement.

The only option, therefore, may be to increase the rate of detection, but this
involves high administration costs that again may be politically unacceptable
both to governments and populations. Even if money is forthcoming, however,
significant problems may be encountered. Take, for example, Northern Ireland.
There is a good deal of evidence that attempts to increase the rate of detection
must confront significant barriers due to the lack of co-operation of the population, especially in the Catholic areas of Belfast (Howe 1990, Leonard 1994).

Indeed, Maguire (1993: 278) argues that 'the use of intimidation by terrorist groups on social security inspectors means that their work cannot be of the same level of efficiency and effectiveness as that of their colleagues on the mainland'. Although such oppositional tendencies to state regulations may not be of the same order of magnitude elsewhere, the important point is that regulations cannot be implemented without compliance from the population. As shown above in the context of Quebec, however, the proportion of the population desiring the eradication of informal employment is not perhaps sufficient to allow this to happen.

A further practical problem with eradicating informal employment concerns the unintended impacts that such a policy may have. The principal reason for eliminating informal employment in the regulatory approach is so that taxes can be raised and fraud/exploitation reduced. In practice, however, this may not occur. Starting with the desire to raise tax revenue, the argument is that informal employment reduces the amount of revenue raised through taxation and that by eradicating it, one can raise more revenue and reduce the burden on the honest tax-payer. To show that one cannot simply assume that all informal employment would become formal employment if such work was eradicated, Mogensen (1985) evaluates the notion that if the 200 million hours of informal employment provided in 1984 in Denmark were converted into jobs for the unemployed, some 110,000 new full-time jobs could be created, thus lowering the unemployment rate of 10.8 per cent in 1984 to 7.8 per cent. He states that this could only be achieved if the extreme price differentials between the formal and informal supply of labour could be abolished. Purchasers of informal employment in his survey, however, state that they would rather resort to do-it-yourself activities (34 per cent) or simply not consume the services (30 per cent) than pay the official formal price. Hence, nearly two-thirds of the work currently undertaken through informal employment would not be converted into formal jobs if informal employment was eradicated. One cannot expect, therefore, informal employment to be fully replaced by formal jobs.[4]

There is also much evidence to suggest that any attempt to eradicate informal employment through tougher rules and regulations will have the opposite effect to that which is intended: it will increase, not decrease, the level of informal employment. To see this, one has only to look at the former Eastern and Central European socialist states. There, state policies aimed at controlling every aspect of economic activity gave rise to a vast range of informal employment (Burawoy and Lukacs 1985, Portes 1994, Sik 1994). This is because, as Peck (1996a: 41) has stated, '*Pressures* for regulation do not necessarily result in effective regulation'. Consequently, and as Portes (1994: 444) puts it,

> The more state policies prevent the satisfaction of individual needs and access to inputs by firms, the wider the scope of informalisation that they encourage. The response will vary, of course, with the specific characteristics of each society. Yet recent evidence on the extent of

irregular activities suggests that state officials have more often than not underestimated the capacity of people to circumvent unwanted rules.

Informal employment can be seen as a constructed response on the part of civil society to unwanted state interference. The universal character of the phenomenon reflects the considerable capacity for resistance in most societies to the exercise of state power. Implementing policy towards informal employment and understanding its consequences thus 'requires a keen sense of the limits of state enforcement and of the ingenuity and reactive capacity of civil society' (Portes 1994: 444).

Indeed, there is a good deal of evidence that the level of informal employment will rise if civil society rejects the principles of the tax and welfare system. An analysis of the 1987 American Taxpayer Opinion Survey by Smith (1992) for instance, reveals that perceived procedural fairness and responsiveness in providing a service were positive incentives that increased tax-payers' commitment to paying taxes. Kinsey (1992), meanwhile, discovers that while detection and punishment attempt to force people to comply with the law, these processes also alienate tax-payers and reduce willingness to comply voluntarily. An increase in the perceived severity of punishment may therefore increase fraud and reduce respect for the fairness of the system. Consequently, in some nations, a simple deterrence model has been superseded. For instance, in New Zealand between 1984 and 1990, tax reforms not only increased the financial penalties but also resorted to media campaigns to persuade tax-payers that their money was being used efficiently on worthwhile projects (Hasseldine and Bebbington 1991). Unfortunately, no evaluation has been conducted of the effectiveness of this strategy.

Raising the level of informal employment, however, is not the only unintended consequence of pursuing tougher regulations. Given that this approach is mostly advocated by those to the left-of-centre, a further unintended consequence, as we shall now see, is that it would exacerbate rather than reduce social and spatial disparities.

Implications for socio-spatial divisions of informal labour

Besides the practical difficulties of eradicating informal employment, a final problem with this approach concerns the implications for socio-spatial inequalities in the advanced economies. In practice, it is much easier and more politically acceptable for state authorities to regulate organised than autonomous informal employment since this is simpler to identify, more exploitative and thus unacceptable to the public, and less deeply embedded in everyday social life. The problem, however, is that this has implications for socio-spatial inequalities. As we have seen in Part II, organised and exploitative informal employment is more likely to be undertaken by marginalised groups such as ethnic minorities, immigrants and the unemployed, whilst autonomous forms of informal employment are more the province of relatively affluent social groups. The impact of any attempt at eradication that sought to eliminate the most exploitative forms of informal employment, therefore, would be to exacerbate existing social inequalities. It would also widen spatial disparities since as shown in a tentative manner in chapter 7, organised informal employment tends to be more prevalent in poorer localities and regions whilst autonomous forms of informal employment are more prevalent in relatively affluent localities and regions.

This also has implications for how such marginalised groups would continue to get by if even these informal coping strategies were taken away from them. Despite the fact that marginalised groups undertake less informal employment than more affluent social groups, this form of work remains a key element for many who find themselves economically marginalised. In southern EU nations, as Reissert (1994) reveals, the proportion of the total unemployed receiving unemployment compensation is very low compared with northern EU nations. Instead, many have to rely on informal employment as a source of income. To attempt to eradicate such employment in these nations which, as aforementioned, also have the weakest welfare provision and formal economies, would leave many bereft of the means to meet their needs and wants, thus leading to further polarisation of living standards in the advanced economies. Indeed, given that women are more likely to engage in organised forms of informal employment and men more likely to conduct autonomous informal employment, it would also exacerbate gender inequalities and result in a patriarchal policy of eradicating a good deal of women's informal employment but maintaining most of men's informal work. Similarly, the outcome would be to deter informal employment amongst ethnic minorities and immigrants by focusing upon eliminating the sorts of work in which they currently engage, whilst leaving much of the work of the dominant white population intact. Such tendencies of increased class, racial, gender and spatial inequalities are but some of the unintended impacts of the pursuit of eradication of informal employment in the advanced economies in practice. There are doubtless many more which could be explored.

Conclusions

Given the way in which Part II showed that informal employment tends to reinforce rather than reduce the socio-spatial disparities produced by formal employment, it might seem superficially that its eradication through tougher regulatory conditions is the most appropriate response. Here, however, the feasibility of such an approach has been evaluated critically. Reviewing the regulatory approach towards informal employment, this chapter has revealed that, despite all of the evidence to the contrary, this approach retains a blind faith in a return to full employment and comprehensive welfare provision. Little consideration is given to the notion that this might not be achievable and thus, that people need to be given alternative ways of engaging in meaningful and productive activity both as a social entitlement and as part of a policy of active citizenship. Examining the feasibility of eradicating informal employment, moreover, this chapter has shown that not only is it impractical given the barriers which confront policy-makers but it is also undesirable due to the fact that it would have the unintended effect of widening socio-spatial disparities.

10

DEREGULATING FORMAL EMPLOYMENT

Introduction

In a market economy without state regulation, there would be no distinction between formal and informal employment. In consequence, and in direct contrast to the approach detailed in chapter 9, some advocate that the only way to rid the economy of informal employment is deregulation. This essentially 'neo-liberal' approach is based on the view that informal employment is a result of the overregulation of the market by the state and also of the dependency culture of welfare induced by unemployment benefits (De Soto 1989, Matthews 1982, Minc 1982, Sauvy 1984). The aim, therefore, is not to promote informal employment as such but to 'formalise' those activities currently undertaken within a relationship of informality by reducing the regulations imposed on employment that force up labour costs and act as a brake on flexibility and thus cause certain businesses and individuals to work informally. Furthermore, the suggestion is made that if the unemployed were given more latitude for self-help, unemployment protection could be reduced, thereby undermining the culture of welfare dependency. Implicit within this approach, however, is a view that the negative impact of such a reduction in unemployment protection on those currently unemployed would be minimal since many are already heavily engaged in informal employment as a principal means of earning a living (De Soto 1989, Matthews 1982, Minc 1982, Sauvy 1984).

In this chapter, therefore, we first outline this deregulatory approach towards work, welfare and informal employment and then evaluate critically the implications of pursuing such a policy approach.

The deregulatory approach to work, welfare and informal employment

Here, the term 'deregulationist' is used to refer to those who argue that informal employment is an indictment of the overregulation of the market by the state and also of the dependency culture of welfare induced by unemployment benefits. Following the structure of the previous chapter, we first examine their view of the future of work and welfare and second, their conceptualisation of informal employment.

The deregulatory view of the future of work and welfare

The most important point to emphasise concerning the future of work within this approach is that its ultimate goal is the return of full 'formal' employment. In this respect, there are no differences between the regulatory and the deregulatory approaches. However, as mentioned above, their explanations of and solutions for the problem of unemployment could not be more different. For these analysts, overregulation of the market is to blame for many of the economic ills befalling society (Amado and Stoffaes 1980, De Soto 1989, Minc 1982, Sauvy 1984, Stoleru 1982). As Peck (1996a: 1) summarises, 'From this viewpoint, failure is seen to have occurred *in* the market, not *because* of the market.' The neo-liberal solution, therefore, is to liberate the labour market from 'external interference' in order to give market forces free reign. Indeed, for many neo-liberal commentators, the growth of unemployment is the result of state regulation frustrating the market in its attempt to cope with the problems of rising productivity and low growth by reducing wages and using labour more flexibly (Minc 1980, 1982, Sauvy 1984, Stoleru 1982).

Sauvy (1984), for example, blames mass unemployment on the rigidity of the economy, expressed in terms of legal controls on employment (e.g. redundancy laws, health and safety legislation, minimum wages), on the social security system and on the general way in which the state imposes itself on the lives of the population. For him, however, legislation in itself is not the cause of rigidity. In an individualistic fashion typical of this approach, he asserts that this legislation is a product of the general desire of individuals for stability and security, resulting in inflexibility. He contends that workers do not seek work as such but a stable job and the fact that jobseekers also seek stable work is a result of, and encouraged by, the welfare system. Sauvy (1984) maintains that it is these attitudes that have led to such legislation and produced a society in which needs go unmet whilst workers claim benefit. Therefore, the way to tackle unemployment is first, to allow wages to fall so that employers will be induced to create jobs and thus absorb surplus labour and second, to cut back on welfare payments so that workers have an incentive to be less fastidious about the jobs which they accept. Consequently, the future of work is one in which market forces hold sway so as to enable supply and demand to return to equilibrium. In this way, the goal of full employment will be achieved.

As witnessed in the UK and US over the past decade or so, the various policies and initiatives that comprise this neo-liberal ideology add up to an economic strategy that places low-waged labour and deregulated labour markets at the forefront of the solution to resolve the 'jobs question'. The implicit and sometimes explicit consequence of such a strategy is to make informal employment less necessary.

So far as the view of the future of welfare is concerned, it is commonly assumed that deregulationists have a very different outlook to the regulatory approach discussed in the previous chapter. However, although this is super-ficially correct, both approaches view the welfare state and the economy as adversaries in that one is usually seen as the root cause of problems in the other. The difference between the two approaches is that whilst the regulatory approach tends to favour the welfare state and views free market capitalism as destroying social equality, deregulationists support the free market and dislike any structure that constrains it. The former, therefore, view the welfare state as a necessary institution for the functioning of modern capitalism and indeed, a prerequisite for efficiency and growth as well as individual self-realisation. The latter, in contrast, consider the nature of the adversarial relationship between the welfare state and economic efficiency to be the opposite. The welfare state is seen to interfere with individual freedoms and the ability of the market to optimise the efficient allocation of scarce resources.

Considering this view of the relationship between welfare and economy, the major debate within the deregulatory approach is over the extent to which a welfare state is required. This debate in neo-liberal thinking is nothing new. As Esping-Anderson (1994) shows, it has been with us ever since the English Poor Law reforms in the early part of the nineteenth century. Within the tradi-tion of classical economics and libertarian thought, one extreme, exemplified by Samuel Smiles, held that virtually any socially guaranteed means of liveli-hood to the able-bodied would pervert work incentives and individual mobility. This, in turn, would stifle the market, freedom and prosperity. This extreme view is echoed in more modern times with some considering welfare provision to be the antithesis of social equality. As social rights are essentially claims against the income and resources of others, these commentators consider the welfare state not as the guarantor of equal status and autonomy for citizens, but as a divisive system under which a class of claimants becomes parasitic upon others' labour and property, with disastrous effects upon their morals (Gray 1984, Murray 1984).

Other more liberal thinkers, such as Adam Smith, however, realised that society did need social provision, especially in health and education. Today, some liberal thinkers are similarly more prepared to accept that the modern economy needs a basic welfare safety net, but stress that due to the conflictual relationship between welfare and economy, this will incur a certain price in terms of economic performance. For these deregulationists, therefore, the issue is where to set the trade-off between equality and efficiency (Barr 1992, Gilder 1981, Lindbeck 1981, Okun 1975).

This spectrum of neo-liberal thought, as displayed above, ranges from those who see no need for a welfare state at one pole to those who see the need for some form of welfare provision for the purpose of equality. All positions on the spectrum, however, emphasise the efficiency trade-offs of pursuing greater equality, particularly with reference to the possibly negative effect of the welfare state on savings (and hence investment), work incentives (and hence productivity and output) and the institutional rigidities that welfare states introduce (such as in the mobility of labour).

Although such internal debates over the degree to which a welfare state should be provided are important to adherents of this approach, the fundamental fact should not be masked that deregulationists are on the whole negative about how the welfare state influences economic performance. For the deregulationists, competitive self-regulatory markets are superior allocation mechanisms from the viewpoint of both efficiency and justice. It follows, therefore, that government interference in the allocative processes (aside from marginal cases of imperfections, externalities or market failure) will risk generating crowding-out effects, maldistribution and inefficiency and the end result will be that the economy will produce less aggregate wealth than if it were left alone (Lindbeck 1981, Okun 1975). Some even go so far as to insist that inequalities must be accepted, and perhaps even encouraged, because their combined disciplinary and motivational effects are the backbone of effort, efficiency and productivity (Gilder 1981).

This deregulatory view of both work and welfare would perhaps be harmless if it was simply an academic theorisation.[1] However, it is not. Despite being a theory that is heavily opposed to state-led change, it is ironically the state in Anglo-Saxon nations, such as the UK and US and to a lesser extent Canada and Australia, which has been the primary vehicle for the implementation of this ideology. Indeed, this is not new. Over forty years ago, Polanyi (1957: 140) recognised that 'the road to the free market was opened and kept open by an enormous increase in continuous, centrally organised, and controlled interventionism'.

From the late 1970s until the change of government in 1997 in the UK, the state rejected Keynesian economic policies of demand management to regulate the economy and committed itself to a radical market philosophy of increasing labour market flexibility, eroding employment rights and removing or privatising welfare protection and regulatory institutions (see Beatson 1995, Crompton et al. 1996, Deakin and Wilkinson 1991/2, Peck 1996a, Pinch 1997, Rubery 1996). The significance of this experiment was not its desire to eliminate or reform outmoded labour market and social institutions and replace them with new institutional structures more suited to current social and economic needs, but rather its explicit policy of fragmenting and dissolving any visible social and labour market institutions which might provide a check on the supposed operation of market forces and the autonomy of firms (Rubery 1996). Hutton (1995) has characterised this as a policy of eliminating all 'intermediary institutions between state and individual' in the interests of driving down labour costs.

In the US, similarly, the process of deregulating formal employment and welfare provision is perhaps even more extreme (see Esping-Anderson 1996,

Myles 1996). In the realm of employment legislation, there has been a slow and continuing whittling away of labour laws, whilst in the realm of welfare policy, there has been a rolling back of welfare provision culminating in the 1996 Welfare Act. According to Peck (1996b), the results of this recent Act are that needs-based funding has ended with funds to individuals being cut off after five years, immigrants have been made ineligible for most federal benefits, and states have been given wide latitude to design workfare and welfare-to-work programmes, being required to cut benefits to those who refuse to work. These workfare-type regimes in the US have been duplicated in other neo-liberal nations, as witnessed in the Jobseekers' Allowance in the UK, introduced in 1996. Similar to the regulatory approach, therefore, the deregulatory approach is much more than an academic exercise. It is being implemented with differing degrees of fervour across many of the advanced economies.

The deregulatory approach to informal employment

For deregulationists, informal employment is seen to be a direct consequence of the encumbrance of state regulations and their costs. Contini (1982) refers to such employment as the 'revenge of the market' for overregulation by the state. Confronted by excessive state regulations, informal workers are thus viewed as heroes (rather than 'villains' as in the regulatory approach) who throw off the shackles of an overburdensome state (e.g. Biggs *et al.* 1988, De Soto 1989, Matthews 1982, Sauvy 1984). In this view, informal employment should not be criticised for undermining employment and short-circuiting the macro-economy of the GNP because it is merely responding, as any other economic activity should do, to 'needs expressed by people' (Sauvy 1984: 274). Informal workers are meeting the needs of society that would otherwise go unmet due to the inevitable way in which state interference in the formal sphere leads to market distortion. Informal employment is thus the people's 'spontaneous and creative response to the state's incapacity to satisfy the basic needs of the impoverished masses' (De Soto 1989: xiv–xv). As Sauvy (1984: 274) explains, informal employment represents 'the oil in the wheels, the infinite adjustment mechanism' in the economy. It is the elastic in the system that allows a snug fit of supply to demand that is the aim of every economy. Indeed, for deregulationists, it is the only mechanism through which one can 'guarantee the utopia of full employment'.

In the view of deregulationists, informal workers are only breaking rules and regulations that are inherently unfair. Such work is thus a form of popular resistance to an unfair and excessively intrusive state and informal workers are a political force that can generate both true democracy and a rational competitive market economy. Given this view of informal employment as one of the last refuges of untrammelled enterprise in an overrigid economic system, these deregulationists see in its supposedly recent expansion a resurgence of the free market against state regulation and union control. According to deregulationists, informal employment is the 'essence of liberalism' (Sauvy 1984). The Italian economist Antonio Martino, for example, considers informal employment 'a masterpiece of my countrymen's [sic] ingenuity, a second Italian economic miracle which has saved the country from bankruptcy, and an example for the other "free" countries to follow' (Martino 1981: 89). Milton Friedman, moreover, asserts that 'the clandestine economy is a real life belt: it effectively limits collective coercion . . . allowing individuals to get round the restrictions imposed by government on personal enterprise' (cited in De Grazia 1982: 480).

In this viewpoint, therefore, informal employment is the locus for the development of pure and perfect competition which is prevented from spreading into the 'modern sector' by the many barriers instituted by the state, such as protectionism, legal measures, excessive bureaucracy, wage rigidity and so on (De Soto 1989). These measures make it possible to maintain barriers to entry that prevent the market from working competitively. In order to avoid these shackles, the (universal) entrepreneurial spirit currently has to operate on the fringes of laws and regulations so as to circumvent the barriers to entry. An example is the set of informal activities known as the 'second economy' in the now defunct state socialist regimes of Central and Eastern Europe. Here, informality became associated with the market and individual freedom. Second economy enterprises producing and trading goods unavailable in state-controlled outlets subverted the logic of the socialist economy in many ways. Even state firms and managers frequently used informal supply sources to overcome the bottlenecks of the official system (Grossman 1989, Lomnitz 1988). Indeed, some Hungarian sociologists assert that the free-market forces unleashed by the second economy were the key solvent that undermined the political legitimacy of state socialist regimes and led to their demise (Borocz 1989, Gabor 1988).

Despite such overwhelming praise from deregulationists for both informal employment and informal workers, this is not to say, however, that the deregulatory approach seeks merely to promote informal employment. That is a common misconception. Rather, its aim is to eradicate informal employment, but not by creating tougher regulations. The intention is to reduce state regulation in the realms of formal work and welfare. This will both unshackle formal employment from the constraints that force up labour costs and put a brake on increasing flexibility, and remove the welfare constraints that act as a disincentive to the unemployed to seek formal jobs. With fewer regulations, the idea is that the distinction between formal and informal employment would wither away so that the two are no longer separate, since all activities would be performed in the manner we now call 'informal', although such activity would be in fact 'formal' since it would not be breaking any rules. In this way, and in this way alone, the goal of full employment is seen to be achievable.

Evaluating the deregulatory approach to work, welfare and informal employment

Having outlined this deregulatory approach towards work, welfare and informal employment, we here evaluate critically the assumptions underlying it as a policy solution to the problem of work and welfare in the contemporary advanced economies and examine its approach towards informal employment.

Critical evaluation of its vision of the future of work and welfare

This approach is based on the idea that full employment can and will return if market forces are allowed to operate unhindered. Here, we assess the validity of such a theorisation. Measured purely in terms of whether this deregulatory approach is more effective in achieving the goal of full employment than the regulatory approach, there is little doubt that this is the case. It does indeed appear that unemployment is lower in advanced economies such as the UK and US that have pursued such a deregulatory strategy with greater fervour than in more social democratic nations of mainland Western Europe. However, their sound performance in terms of job creation must be seen both in terms of the quality of the jobs created and the degree of social polarisation which this 'success' has entailed (see Conroy 1996, Esping-Anderson 1996, European Commission 1996b, Fainstein 1996, OECD 1993, Peck 1996a and b, Rubery 1996).

In the advanced economies, two distinct approaches to social polarisation have been adopted (see Pinch 1994, Williams and Windebank 1995b). First, and most widely used, is the approach that examines inequalities concerning employees in the employment-place, such as in terms of widening income or earnings differentials. Second, and popular in the UK, is the approach that takes the household as the unit of analysis and examines the increasing disparities in household income. Whichever approach is adopted towards social polarisation, the finding is the same. The neo-liberal nations appear to be witnessing social polarisation to a much greater degree than the more social democratic nations. So far as the former approach is concerned, the OECD (1993) reports that during the 1980s in neo-liberal nations, the lowest-decile earners lost ground relative to the median, by 11 per cent in the US, 14 per cent in the UK, 9 per cent in Canada and 5 per cent in Australia. For example, and as Fainstein (1996) shows in the US, the percentage of permanent full-time workers with earnings below the official poverty line (about $13,000 in 1992) increased by 50 per cent from 1979 to 1992 from 12.1 per cent of workers in 1979/80 to 18.0 per cent by 1992/3. Such disparities, moreover, are far higher than in social democratic nations. Workers in the lowest earnings decile in deregulationist nations such as Canada and the US, for example, earn about 40 per cent of median earnings compared with 70 per cent or more in social democratic nations such as Germany and in Scandinavia (OECD 1993).

Examining the latter approach, based on widening household inequalities, moreover, the same distinction between neo-liberal and social democratic nations in terms of the degree of social polarisation is identified. As Myles (1996) reports of the US, between 1973 and 1987, the income of the richest 20 per cent of families increased by 25 per cent whilst that of the poorest 20 per cent fell by 22 per cent. More widely, examining Table 9.1 in the previous chapter, a clear trend is apparent. The US and the UK, as the two nations that have led the race towards deregulation, are also the nations with the highest levels of social polarisation between households. These nations have the highest proportion of households that are either multiple-earner or no-earner households. We can thus only agree with Peck (1996a: 2) that, 'Contrary to the nostrums of neo-liberal ideology and neo-classical economics, the hidden hand of the market is not an even hand.'

A further problem is that there is little evidence that some of the fundamental tenets of this ideology will have the impact deregulationists desire. Take, for example, the policy of stripping away the welfare state so as to encourage people to find employment. Numerous studies of the effects of reducing welfare benefits on the levels of unemployment conclude that decreasing levels of benefit (or withholding benefit) will not cause an increase in flows off the unemployment register (Atkinson and Micklewright 1991, Dawes 1993, Deakin and Wilkinson 1991/2, Dilnot 1992, Evason and Woods 1995, McLaughlin 1994). In an extensive review of the effect of benefits on (un)employment, McLaughlin (1994) concludes that the level of unemployment benefit does have some impact on the duration of individuals' unemployment spells, but the effect is a rather small one. Following Atkinson and Micklewright (1991) and Dilnot (1992), she states that the level of unemployment benefits in the UK could not be said to contribute to an explanation of unemployment to a degree that is useful when considering policy. Moreover, extremely far-reaching cuts would be required in benefit levels to have any significant impact on the duration and level of unemployment. The effect of such cuts would be to create a regime so different from the present one from which the estimates of elasticities (of unemployment duration with respect to out-of-work benefits) are derived, that their predictive usefulness would be very suspect (Dilnot 1992).

Neither will the taking away of the cushion of the welfare state simply allow people to get by on their own. The dependency culture resulting from the onslaught of formal welfarism has stripped away the institutions of civil society that once allowed people to survive outside the nexus of the welfare state (see, for example, Mingione 1991). Therefore, the possibility that greater self-help will occur simply by taking away people's 'safety net' appears mistaken. The net result is more likely to be that people will be simply left bereft of the means of survival. Nevertheless, given the increasing pressures being put on formal welfare provision in the advanced economies, it remains obvious that something will need to be done about the way in which welfare is provided. The choice, however, is not either to strive to restore full employment and a comprehensive welfare 'safety net' for the economically excluded as the regulationists advocate, or to strip away the welfare 'safety net' and deregulate the formal labour market as a means of encouraging full employment, as the deregulationists propose. As we have shown, the former is impractical and flies in the face of the direction of the advanced economies, whilst the latter, even if it were to achieve full employment, would do so at an extremely high price in terms of the levels of absolute and relative poverty. Instead, it will be proposed that there exists a middle path that seeks full engagement (rather than full employment) and which provides welfare support to those who need it whilst actively giving people the means to help them help themselves. This alternative approach towards the future of work and welfare will be returned to in the next chapter. For the moment, we must turn our attention towards an evaluation of the deregulationists' view of informal employment.

Critical evaluation of its view of informal employment

For many deregulationists, informal employment is seen to be not only an indicator of the way forward for employment but a means by which the unemployed and marginalised are currently getting by in the advanced economies. As such, many assert that there is little need for concern over the dismantling of the welfare 'safety net' (e.g. Matthews 1983). The problem, however, and as Part II of this book has already shown, is that this is not the case. Not only do the unemployed, for instance, engage in informal employment to a lesser extent than the employed, but such a form of work seldom provides sufficient income to enable poorer populations to get by. Such employment is nowhere found to be an adequate substitute for a formal job and formal welfare provision. Consequently, there are good reasons to suppose that if the formal labour market and formal welfare provision were further deregulated, then social and spatial inequalities would become more exaggerated.

Reviewing the way in which a range of economic, social, institutional and environmental factors usually combine in most localities in the advanced economies to prevent the poor and unemployed from engaging in informal employment (see chapter 4), we suggest that there is no reason to believe that deregulation of formal welfare provision will improve their participation in formal or informal employment. For example, the reason why many of the poor and unemployed undertake little informal employment is because they frequently lack the money, tools, social networks, skills and opportunities to engage in such work (Pahl 1984, Williams and Windebank 1993, 1994, 1995a and b, 1997, Windebank and Williams 1995, 1997). Labour market deregulation will have no impact on these factors: it would not improve their skills or their social networks. Neither would it provide them with access to the tools and money necessary to engage in this form of employment. All that such a move would do is to enhance the incentive of the unemployed to work informally in order to survive.

This alone, however, appears to be insufficient to tackle the barriers that prevent those excluded from the formal labour market from even participating in informal employment. As Gilbert (1994: 616) argues, 'The hope that it [informal employment] can generate economic growth on its own, that the micro-entrepreneurs can go it alone, with a bit of credit and some deregulation, seems to be hopelessly optimistic.' Therefore, in adopting a deregulatory approach which, for example, asserts that the way to solve the problem of poverty and unemployment is to take away formal welfare support, with little if any state assistance to facilitate the transition, the only possible reason that greater economic competitiveness and socio-economic equality would be achieved is because there would be a levelling down, rather than up, of material and social circumstances for the vast majority of the population.

It is not only the case that deregulation seems insufficient as a tool for aiding the poor and unemployed. As evidence of how deregulation of formal employment and the welfare system is also insufficient by itself to enable poorer localities and regions to improve their circumstances, there are several lessons to be learned from Emilia Romagna in Italy. As Amin (1994) makes clear, the success of this region in turning itself into a competitive area is founded not only on the deregulation of the formal sphere but equally, on strong public sector support and co-ordination. The success of Emilia Romagna has been very much dependent upon strong state intervention such as industry-specific services to firms to foster task specialisation and interfirm co-operation to advertise the products of an area and to secure the long-term reproduction of sector-specific skills. This view that the success of this area is based on 'controlled deregulation' is reinforced by many other studies (e.g. Cappechi 1989, Warren 1994). These all find that the creation of contemporary Marshallian industrial districts in this region are not only the result of a particular 'cocktail' of conditions being present (e.g. in interfirm co-operation, structure of sociability, local 'industrial atmosphere' and 'institutional thickness') which have allowed this transition to occur, but are also due to strong state intervention in co-ordinating this transformation.

Consequently, not all areas can replicate the success of Emilia Romagna by pursuing a policy of controlled deregulation alone. This area has a distinctive 'cocktail' of pre-existing conditions that have allowed it to flourish, which cannot necessarily or easily be reproduced elsewhere. Deregulation in other environments not possessing such a local industrial atmosphere, interfirm co-operation and structure of sociability would have very different results. On its own, therefore, deregulation is insufficient to make poorer regions and localities more economically competitive. Furthermore, and as we have seen, systemic support is required for firms along and across value-added chains. It is clear that deregulation on its own will not revitalise deprived regions and localities.

Conclusions

This chapter has revealed that the deregulationist approach, as with the regulatory approach in the previous chapter, desires a return to 'full employment'; but the nature of the 'full employment' economy envisaged by the former is very different from that of the latter, as is the means of achieving this end. Rather than bolstering what they see as the flagging edifices of 'diluted' or 'welfare' capitalism through state-led work-sharing and job creation as well as comprehensive welfare support, supporters of this approach advocate 'undiluted' capitalism. The intention, in so doing, is to close the gap between formality and informality by eradicating the regulations and constraints that currently hinder the achievement of full employment. Consequently, although many exponents of this approach see the already deregulated sphere of informal employment as an exemplar of economic and social organisation, they do not wish to promote or protect it in its present form. In sum, this approach seeks to replace the formal/informal dichotomy not by regulating informal employment out of existence but by deregulating formal employment to a degree that renders informal employment unnecessary.

Evaluating critically the implications of pursuing this approach, this chapter has revealed that although there is little doubt that deregulation of formal employment would reduce the magnitude of informality, since informal employment is by definition a product of the regulations imposed on formal employment, it is very doubtful that deregulation of formal employment and the welfare state would enable the socially excluded to improve their circumstances. On the one hand, it has revealed that the impact of such deregulation has seen social inequalities widen at a quicker rate in those nations that have pursued deregulation than in more social democratic countries. On the other hand, it has shown that deregulation will not automatically allow marginalised groups and areas to solve their problems.

11

INFORMAL EMPLOYMENT
AND THE NEW ECONOMICS

Introduction

Like the deregulationist approach, this final approach views the existence of informal employment as an indictment of the way in which economic life is currently organised. However, whereas the deregulationists criticise the over-regulation of the economy and welfare provision by the state, the advocates of the 'new economics' disagree with the current arrangements for distributing income and work. The aim of this approach, therefore, is to seek a radical restructuring of work and welfare that will abolish the necessity for people to undertake exploitative informal employment.

Although the new economics is an intellectual home to a wide-ranging mix of analysts, all have a number of common strands running through their work. The journal of the New Economics Foundation, *New Economics*, defines these as providing 'economic justice; satisfaction of human needs; welfare for people and planet; satisfying and democratic work; self-reliance and self-determination; encouragement for informal economic activities; and respect for the present and future inhabitants of the earth'.[1] Therefore, this approach is supportive of developing informal economic activity as a means of promoting self-reliance and self-determination, but not in the same way as the deregulationists. These analysts assert that it would be a grave mistake to associate the new economics with either right-wing political thought (e.g. the deregulatory perspective towards informal employment) or left-wing thinking (e.g. the regulatory approach towards informal employment). For them, both left- and right-wing beliefs describe different ways of improving productivism and engendering full employment. The new economics, in contrast, seeks to replace the employment-focused productivist mentality of the other approaches with a 'whole economy' view of work. This pursues 'full engagement' (e.g. Lipietz 1995, Mayo 1996, OECD 1996) based on the active promotion not only of formal employment but also unpaid informal work.

In the first section, therefore, we highlight the new economics approach towards both work and welfare in general and informal employment in particular. We then evaluate critically the potential problems in implementing such an

158

approach. As in previous chapters, this reveals that the options open to policy-makers concerning informal employment are part of the wider debate on the future of work and welfare in the advanced economies.

The new economics approach to informal employment

In both the regulatory and deregulatory approaches, there is a common assumption that full employment can and should return. The principal differ-ence between them is that they have contrasting ideas over how this can best be achieved. The new economics approach, however, works on the basis that the age of full employment and its attendant comprehensive welfare state is over and that this is not a passing to be mourned. On the issue of why full employment and its attendant welfare state cannot return, the new economics draws upon the type of evidence used to evaluate the regulatory approach in chapter 9. Full employment, if it ever existed, is seen to have done so in just a few advanced economies during the middle of the twentieth century and is viewed as a way of organising work which is becoming outdated (Gass 1996, Mayo 1996, Rifkin 1995, Robertson 1985, 1991). To seek its return is merely to engage in mistaken and unrealistic 'golden age' thinking where the future is envisaged simply as a return to some previous era.

For these analysts, however, it is not simply the case that full employment will not return. It should not return. Three principal reasons are given for this prescription. First, it is maintained that most people are unsatisfied with their formal employment because their jobs are often stultifying, alienating and do not allow them the freedom to compensate for this lack of satisfaction outside working hours (Aznar 1981, Gorz 1985, Laville 1995, 1996, Mayo 1996, Robertson 1985). Given this lack of opportunity for personal growth in employment as well as the fact that the only alternative to a job is unemploy-ment which cannot provide self-esteem, social respect, self-identity, com-panionship and time structure, the current age of employment is not perceived as at all positive for these new economists. Aznar (1981: 39) assesses this situa-tion in the following manner: 'any society which proposes that its citizens spend the whole of their time, energy and empathy engaged in an activity which cannot, by its very nature, soak up this energy, is fundamentally perverse'.

The second reason why employment should not remain centre-stage flows from their observation that society, in pursuing economic growth for its own sake rather than as a means to an end, has lost its way (Douthwaite 1996, Gorz 1985, Mayo 1996, Robertson 1985). In consequence, an individual's work, as embodied in employment, no longer responds to the real needs of the con-sumer. The result, as Friedmann (1982) states, is that only 35–40 per cent of the economically active population are engaged in 'indispensable' produc-tion; the rest are obliged to produce essentially unnecessary goods in order to earn the income necessary for personal survival. Gorz (1985: 58) agrees and suggests that 'for a section of the population, only the production of inessentials allows them access to necessities'. For the new economists, however, nobody should have to work at such production to earn the money necessary for survival: 'employment should be seen not as an end in itself, but as a means to achieving a better quality of life' (Mayo 1996: 147).

Third and finally, these new economists argue that the relentless pursuit of full employment has led to a devaluation of unpaid informal economic activity. The prolonged structural crisis of unemployment, however, is considered to provide an opportunity to revalue this form of work. As the recent OECD (1996) report produced by some of these analysts suggests, there is a need to put the economy back into society rather than see it as an autonomous or independent element. For them, and mirroring much of the recent academic interest shown in analysts such as Polanyi and Granovetter, the desire is to develop a socially embedded view of economic activity (see, for example, Lee 1996, Verschave 1996). The crisis of unemployment, therefore, provides an impetus for just such a rethinking about the justification of subjugating all social goals to the economic aim of continued growth rather than viewing the 'economy' as serving the interests of society.

Considering this view that full employment cannot and should not return, these new economists advocate a different organisation of work and welfare. Their argument is that there is a need and desire to shift towards more mixed economies of production and welfare; that is, economies that promote both unpaid informal work and formal employment. Nevertheless, these analysts do not argue for harnessing informal employment *per se*. They too wish to replace informal employment. In contrast to the previous two approaches, however, they argue that there is nothing to be gained by increasing the controls on informal employment since attacking these symptoms of the inherent problems of our economic system will not cure them. Nor do these analysts seek the *de facto* officialisation of informal employment by abandoning laws protecting formal workers and cutting back on social security, since they believe this would exacerbate the social problems that the present organisation of work and welfare provoke (see Sachs 1984). Instead, they believe in assisting people to harness their productive energies so that they can help themselves. Although superficially this may appear similar to the proposal of the deregulationists for greater self-help, this assisted self-help approach differs on three counts. First, it is based upon the logic of supplementing and not substituting formal activity and state provision. Second, it is based on the concept of optionality and choice, which contradicts the conservative appeal to duties and norms. Third and finally, such informal provision is envisaged more in terms of collective and interactive forms of working instead of in terms of isolation and competition.

Related to this pursuit of assisted self-help is the belief that we need to shift the welfare debate away from the either/or discussion of 'rights' (e.g. European Commission 1996b) or 'responsibilities' (e.g. Etzioni 1993) and towards an approach which includes both. As Ekins (1992) asserts, for example, everyone has a right to paid work as well as access to the necessary resources (e.g. skills, land, workshops) to enable them to do unpaid work, but everyone also has a responsibility to work according to their abilities both to generate wealth corresponding to the goods and services they consume, and to contribute to the society which guarantees them, among other things, the right to work. Embedded in this synthesis of the rights versus responsibilities debate, therefore, is a deeply ingrained notion of active citizenship as opposed to the enforced idleness wrought by the current dependency culture of welfare provision.

To achieve this new organisation of work and welfare, the most frequent call is for a two-pronged approach. On the one hand, the new economists advocate a top-down state-led solution similar to the regulatory approach. However, based on the assumption that full employment and a comprehensive welfare state will not and should not return, their idea is to introduce a 'basic income scheme' as a means of delinking income from employment and creating a new welfare 'safety net'. On the other hand, the new economists couple this with a grass-roots solution towards work and welfare. Although this is similar to the deregulatory approach in the sense that it pursues self-help, the difference is that it does not see this as a way of returning to full employment, and much greater emphasis is put on assisting people to help themselves rather than simply casting them adrift. Here, therefore, we examine each in turn.

The top-down solution: a basic income scheme

Questioning the supposedly inextricable relationship between employment and income and recognising that less than two-thirds of the population of working age in many advanced economies now have a job, the new economists have asked whether employment is the most appropriate means by which income can and should be distributed. Given that only just over half of all income now derives directly from employment (Rosewell 1996), they argue that there is a need to reconsider fundamentally the supposedly inextricable relationship between employment and income.

Their view is that unless there is a radical restructuring of work and income, the result will be both the increasing impoverishment of a larger proportion of the population as employment becomes more concentrated in an ever more limited range of households (see Table 9.1 earlier) and the spread of badly-paid insecure jobs on the fringes of the formal labour market, including exploitative informal employment (e.g. Gorz 1985, Offe and Heinze 1992). To prevent this, their proposal is to introduce a 'guaranteed basic income scheme' (Daly and Cobb 1989, Duchrow 1995, Lipietz 1995, Mayo 1996, Robertson 1994, 1996). Alternatively known as a social wage, social dividend, social credit, guaranteed income, citizen's wage, citizenship income, existence income or universal grant, within such a system every citizen would receive a basic 'wage' as a social entitlement without means-test or work requirement. With this minimum income guarantee in hand, individuals could then choose whether they wish to improve their well-being by engaging in employment so as to earn additional money in order to purchase goods and services, or whether they wish instead to invest their time in self-provisioning those goods and services. By introducing such a system of income allocation, the belief is that those presently unemployed would be freed to engage in activity, whether paid or unpaid, to improve their own well-being and as an outlet for their creative potentials and/or entrepreneurial spirit, and the 'poverty trap' would be abolished (see Jordan and Redley 1994, Mayo 1996, Parker 1989, Van Parijs 1992, 1996a and b). Exploitative informal employment, moreover, would become a thing of the past because nobody would be obliged to engage in such work to augment their incomes to a basic survival level (e.g. Jordan and Redley 1994, Mayo 1996). As Mayo (1996: 158) asserts,

> The approach of a Citizen's Income . . . would help to support bene-
> ficial changes in working patterns and practices by removing un-
> employment and poverty traps, eliminating today's black economy of
> undeclared earnings, raising pay levels for dirty and unsociable jobs,
> and encouraging work in the [unpaid] informal economy.

Even amongst advocates of a basic income, however, it is now accepted that a fully individualised and unconditional basic income could not be introduced in one operation, if only because of the way in which it would upset the current distribution of incomes and labour supply. Instead, and particularly for the working-age population, there is a growing consensus that one should not pro-ceed by gradually introducing a full basic income on a group-by-group basis, but start with a very modest (partial) basic income that would not be a full sub-stitute for existing guaranteed minimum income provisions (when they exist). In several nations, serious costing exercises are underway. These have raised a series of questions. Could and should the introduction of a partial basic income be accompanied by a matching cut both in other benefits and gross wages, by reductions in other benefits only and higher taxes on wages, or by a higher uni-form taxation of other benefits and labour income? Should the basic income be added to the taxable income? Should current tax exemptions on the lower range of labour income be kept as they are, reduced or abolished? Is there any short-term prospect for alternative funding (e.g. generalised social security con-tribution, energy tax, VAT, gross profits of enterprises)? Should the level of the partial basic income be explicitly indexed to the cost of living, to the minimum wage or to GDP per capita? (Citizen's Income Trust 1995). Such questions are the source of a great deal of debate within the citizen's income literature and some of the issues arising will be returned to below (Atkinson 1995, Dore 1996, Meade 1995, 1996, Parker 1989, Van Parijs 1996a and b).

Here, however, it is simply important to recognise that for many involved in the new economics, such a top-down solution, although necessary, is by itself insufficient as a tool for restructuring work and welfare and eliminating informal employment. Instead, this top-down solution needs to be coupled with initiatives to encourage 'active citizenship'. As Lipietz (1992: 99) puts it, his proposal for a universal basic allowance of two-thirds the minimum wage 'would be acceptable only if it meant that those who received it were prepared to show their solidarity with society, which is paying them'. For him, this requires the creation of a new sector engaged in socially useful activity and comprised of 10 per cent of the labour force (the unemployment rate at the time he was writing), who would receive a normal wage for their work. Most new economists would agree with the sentiment that a basic income needs to be coupled with active citizenship, but would disagree with Lipietz about how this should be achieved. For most new economists, it is necessary to couple the top-down solution of a basic income with the bottom-up solution of assisting people to help themselves rather than creating yet another top-down imposed structure to encourage active citizenship.

The bottom-up solution: assisted self-help

For new economists, assisted self-help is a necessary complement to a basic income scheme so as to ensure both that the 'dependency culture' which has been the outcome of the formal welfare state is not replicated and that the socio-spatial inequalities in informal work are not preserved. For them, the intention is not to 'colonise' (Heinze and Olk 1982) informal economic activity by incorporating it into the formal wage system as in the deregulatory approach and neither is it to put greater emphasis on promoting formal employment without regard for informal work as in the regulatory approach. Rather, the intention is to pursue the promotion of formal employment and at the same time, harness unpaid informal work so as to enable those marginalised from employment to develop the means to defend themselves and survive. This approach towards unpaid informal work, variously referred to as 'liberal self-help' (Evers and Wintersberger 1988), a 'complementary' approach to local economic development (Macfarlane 1997), a 'complementary network' model (Heinze and Olk 1982), 'DIY citizenship' (Simey 1996) and a 'social economy' approach (Porritt 1996), thus aims to use self-help to 'level the playing field' and empower disenfranchised individuals and communities through capacity-building to help themselves. By reversing the low status and conditions of unpaid informal work, the intention is to replace what is seen as 'useless toil' (much of the work currently undertaken in employment) with 'useful work' and in so doing, to move towards 'full engagement' whereby all citizens have access to income and to meaningful work, paid or unpaid (Mayo 1996).

Within this bottom-up approach of the new economics, two strands can be identified: 'ecologists' and 'social democrats'. The ecologists (e.g. Desproges 1996, Ekins 1992, Mayo 1996, Robertson 1985) advocate this bottom-up solution not only as a means of improving people's quality of life but because diversified and decentralised activity is believed to be much more in harmony with ecological principles and sustainable development. For them, encouraging a whole economy approach is necessary to avoid environmental disaster as well as to provide a more fulfilling life for individuals. Social democrats, meanwhile, pay less attention to environmental issues and place more emphasis on promoting a more flexible organisation of working time which frees employees from the stifling effects of work organised solely around employment, as well as liberating the unemployed from a dependency culture (Aznar 1981, Greffe 1981, Laville 1996, Nyssens 1996, Roustang 1987, Sachs 1984). This bottom-up strategy, the social democrats believe, will provide a flexible and innovative antidote to recession and the problem of mass unemployment. However, unlike the ecologists, they do not believe that such a whole-economy approach necessitates either a reduction in formal work or a questioning of economic growth. Both strands, nevertheless, agree that the formalisation of economic life has resulted in an increasing inability of unpaid informal economic activity to meet economic and social needs. As Robertson (1981: 89) asserts, until now, 'no one has stopped to ask whether people would be better off if we generally lived a greater proportion of our lives in the [unpaid] informal economy, or what would be the right balance between the two halves of the dual economy'.

How, therefore, will such an assisted self-help approach be implemented? For the new economists, many of the solutions already exist. They point to a whole range of grass-roots initiatives around issues such as credit (e.g. credit unions), money (e.g. local currency systems such as LETS and time dollars) and agriculture (e.g. local food links) as well as the shift towards participative planning and research, neighbourhood involvement, decentralisation of services and bootstrap community economic initiatives (see, for example, Douthwaite 1996, Williams 1997). Here, we take just one example of a grass-roots self-help initiative to reveal the ways in which such informal solutions can assist self-provisioning.

Assisting people to help themselves: the case of Local Exchange and Trading Systems (LETS)

LETS are local associations whose members trade goods and services with each other using a local unit of currency (e.g. acorns in Totnes, favours in Calderdale, solents in Southampton). To do this, each member lists in a directory the range of goods and services which they would like to offer and receive. Individuals then decide what they want to trade, whom they wish to trade with and how much trading they wish to do. The price, moreover, is arrived at in mutual agreement between the purchaser and seller, although the seller will sometimes state the price s/he wishes to receive in the directory. No cash, however, is issued. All transactions are recorded by means of cheques written in the local LETS unit. The association maintains an account of these transactions in the same way as a bank. The cheques are filed by a treasurer who sends out regular statements of account to the members. No interest, moreover, is charged or paid and credit is freely available. One does not, in other words, have to sell anything before purchasing.

Although this phenomenon has historical roots (see Pacione 1997), the contemporary origins are in the early 1980s when Michael Linton set up such a system in Courtenay, British Colombia. The idea reached the UK in 1985 when he presented a paper at The Other Economic Summit (TOES) (Linton 1986), a forum for new economics thinkers which runs alongside the G7 Economic Summit. From that point onwards, it has been advocated and developed in the UK and elsewhere, mostly by those involved in the new economics (Brandt 1995, Dauncey 1988, Dobson 1993, Greco 1994, Lang 1994). Indeed, as recent surveys reveal, the origins of LETS in the UK, Australia and New Zealand can all be traced back to that 1985 TOES workshop (see Williams 1996a, b, c, d and e). It is only in the 1990s, however, that LETS have become more widespread. In early 1992, for instance, there were just five LETS in operation in the UK but by late 1994, this had grown to more than 200 with over 20,000 members and by mid-1995, there were over 350 LETS in the UK with more than 30,000 members in total trading over £2.1 million per annum (Williams 1996d).

Therefore, do LETS help people to help themselves? As Lang (1994:2) puts it, 'there is much work in our world needing to be done, many people willing and able to do it, but seemingly insufficient money to pay them'. LETS overcome this by providing the money to enable people to engage in self-help. Take, for example, the unemployed who comprise 29.4 per cent of LETS members (Williams 1996d). For this social group, LETS can in theory transcend all of the constraints discussed in chapter 4 that prevent them from engaging in informal work in general, and informal employment in particular. Firstly, there is the fact that redundancy leads to a reduction in the size of their social networks, meaning not only fewer chances of hearing about opportunities for informal work but also fewer offers of help (Miles 1983, Renooy 1990, Morris 1993, Thomas 1992). In theory, LETS can counter this tendency by providing a means of bolstering the size and diversity of their contacts, which helps surmount the problem of social exclusion. Such networks may also give the unemployed back some of their self-esteem by empowering them to participate in work and to demonstrate their worth through their economic activity. Indeed, in surveys conducted in Manchester, Totnes and Calderdale, around three-quarters of respondents state that the LETS has bolstered their social networks, provided them with a source of activity and increased the range of people that they can turn to for help (Williams 1996a, b, e and f).

Second, the unemployed do not have the money to acquire the materials necessary to undertake informal work (Pahl 1984, Smith 1986, Thomas 1992). LETS, however, mean that money is no longer a restriction. Those unable to buy materials can acquire or borrow them from LETS members. Such schemes can also provide them with access to credit that is usually denied them

(Leyshon and Thrift 1994). The evidence from the surveys is that this is indeed occurring. Many members use LETS as a means of interest-free credit and as a way of getting access to materials and resources (e.g. ladders, DIY equipment) to do jobs for themselves. Third, the unemployed lack the requisite skills to tackle some informal work (Smith 1986). For example, if their skills are inappropriate for securing a formal job, it is unlikely that they will be in demand informally (Williams and Windebank 1995a, 1997). This problem is overcome on LETS, however, especially in those systems which count all activities undertaken as being of equal value, introduce a minimum wage rate or develop a training aspect so that skills can be acquired which could be used both on the LETS and sometimes later, to gain access to formal employment (Williams 1996e).

Finally, the unemployed undertake less informal employment because they are more inhibited by fear of being reported to the social security authorities. At present, although there is some confusion concerning whether or not it is legal for the unemployed to earn LETS currency and, if so, how much they can earn, LETS nevertheless represent a means for the unemployed to improve their circumstances in a regulated environment, thus lessening the probability of turning to informal employment or crime. Indeed, LETS are formalising work which would otherwise take place through informal employment. The studies conducted in the UK by Williams (1996a, b, d, e and f) suggest that some 21 per cent of the work undertaken on LETS would have been conducted through informal employment if the LETS did not exist. LETS, therefore, have the ability to shift much of what is currently undertaken as informal employment into a more regulated environment, albeit one that is regulated by LETS members themselves rather than by the state.

Consequently, LETS can directly attack each and every one of the barriers to the participation of the unemployed in informal work. There is also mounting evidence that they enable members to pursue productive and meaningful work that they find satisfying. The resounding finding of the limited number of studies so far conducted is that members engage in activities on LETS through which they wish to define themselves and seek social identity. For example, formally employed accountants weave baskets, unemployed women provide aromatherapy and unemployed men provide gardening or home brewing services (Williams 1996a). In other words, members see LETS as a way of engaging in self-satisfying 'useful work', something that is denied them in their formal job or life as a social security claimant. Besides providing access to meaningful, productive and satisfying work both for the unemployed and those engaged in stultifying employment, there is also evidence that LETS are by definition an effective and efficient tool for pursuing sustainable development. Not only do they reduce transport costs and pollution by encouraging people to engage in local exchange, but LETS also facilitate greater recycling, reuse and repair of goods that would otherwise be thrown away. As such, many in the new economics see them as fulfilling their desire for sustainable development (e.g. Barnes *et al.* 1996).

LETS, however, is but one grass-roots initiative that these new economists seek to promote in order to help people to help themselves and in particular, to help the economically excluded become more self-reliant. Indeed, there is currently much work being undertaken on how grass-roots initiatives such as LETS, credit unions and local food link projects can be integrated so as to create the synergies necessary if people are to be enabled to help themselves more effectively. Having outlined the new economics approach towards work and welfare in general and informal employment in particular, attention now turns towards evaluating this approach.

Evaluating the new economics approach to informal employment

Here, therefore, we analyse the practicality and desirability of implementing the new economics approach towards informal employment. To do this, we commence by evaluating the top-down solution of introducing a basic income scheme and then the grass-roots solution of assisting people to help themselves.

So far as the practicality and desirability of implementing a basic income scheme in the advanced economies is concerned, first, there is the issue of the cost. Superficially, one might consider that the transfer of the current costs of the vast array of welfare expenditures and tax-free allowances could finance a basic income scheme. However, in practice, even exponents of this approach agree that this would be inadequate to provide everybody with a full basic income (i.e. enough to live on) and assert that such high levels of income tax (probably around 70 per cent) would be required that it would be unlikely to gain political or public support (Citizens Income Trust 1995). Due to such problems, many supporters of a basic income scheme discuss only the feasibility of either a partial basic income which would be at about half the rate of income support for a married couple with less for children and pensioners, or a transitional basic income scheme which is yet another variant and seeks to introduce basic income in stages.

Besides the cost issue, another problem relates to the 'political will' to implement such a scheme. First, this is due to opposition to such an approach from all the institutions and social forces whose existence depends on the present system of production, distribution and consumption, such as large corporations (see Heinze and Olk 1982). Second, it is due to the widespread fear that detaching income from employment would reinforce the current trend whereby consumption is replacing production as a major vehicle for self-identity and source of social interaction (e.g. Glennie and Thrift 1992, Marsden and Wrigley 1994), meaning people will not wish to participate either in employment in particular, or active citizenship more generally. However, given that

167

to seek self-identity through consumption, one needs an adequate level of income, there is little reason to believe that the level of basic income would act as a catalyst for such a shift. Third and finally, it is because there is a question mark over whether the redistribution that such a scheme entails would be acceptable in the present political and economic climate.

A further problem concerns the impact of a basic income scheme on informal employment. One unintended implication is that it could well lead to a rise in informal employment. This is because presently, there are many people in countries such as the UK earning too little to pay tax and who are thus not defined as engaged in informal employment. This would not be the case under a basic income scheme. Presumably, if tax were paid on any formal employment undertaken, work which currently falls outside the definition of informal employment would become informal. For example, where a homemaker engaged in a limited amount of paid work for which s/he would not be liable for tax due to the low amounts of income earned under the current system, s/he would not be engaged in informal employment, but under a basic income scheme, if any activity undertaken was taxed, s/he would now be undertaking such employment informally. A further reason why informal employment might increase under a basic income scheme is because currently a major dis-incentive preventing the unemployed from engaging in such work in many advanced economies is that their social security payments are taken away from them if they are caught working whilst claiming. If a basic income was a right and/or entitlement, then this would no longer apply. For some, therefore, this would mean that they could engage in informal employment without fear of the consequences. For others, however, this basic income would liberate them to give up their informal employment and augment their income with a formal job. Which direction it takes would be in major part dependent upon the graduation of tax. The more graduated the tax, the more informal employment would become formalised. The steeper the graduation, in contrast, the more people would work informally.

Another impact is that although a full basic income scheme would be likely to eliminate the most exploitative forms of informal employment since it would take away the necessity for most people to engage in such activity in order to earn an income to get by, it might well lead to a surge in other forms of infor-mal employment. This is because it could increase the demand for organised informal employment amongst employers confronted with very high tax rates and would certainly lead to an increase in both the demand and supply of autonomous informal employment. A partial rather than full basic income scheme, meanwhile, is unlikely even to erase the exploitative forms of informal employment since the incomes would be inadequate for people who currently have to turn to informal employment to survive. As Lipietz (1992) asserts, if the level of partial basic income is set too low to allow a dignified standard of living, it then becomes a kind of subsidy for formal and informal employers who, according to need, can take on people from a pool of unemployed just kept at subsistence level. If it is too high, however, then it might deter people from working, except for very highly-paid people and those who work for other ends besides income, such as social status, networks and social contact. Indeed, this last point is important. Its impact on those forms of informal employment conducted more for social than economic reasons would be slight.

There are also issues concerning the impact of a basic income on the work patterns of immigrants. A so far neglected problem with a basic income scheme is that it is only guaranteed to citizens. As such, it does not deal with the case of illegal immigrants who currently engage in informal employment. For them, little would change. Indeed, the rate of illegal immigration might well increase given the rise in wage rates of those citizens guaranteed their basic income, since the demand from employers for non-citizen labour might rise. It might also be the case, depending on the immigration laws in place, that not only illegal immigrants but other categories of immigrant would also be excluded from such a citizen's income. The effect would be to extend their participation in informal employment. A basic income scheme, therefore, might reinforce the current racial divisions of labour. In sum, it appears that a basic income scheme not only runs into problems concerning the practicality of implementing it but might also be undesirable if implemented alone because it would be insufficient to stem the prevalence of informal employment and could well lead to the widening of some social disparities.

As can be seen from the above, there are many complications in thinking through the problems of implementing a basic income. What is perhaps required, therefore, before going any further, is for pilot projects of various types of basic income scheme to be run so as to evaluate their impact on economic and social life. As this book has highlighted, these will need to be conducted in different localities with varying economic, social, institutional and environmental characteristics, since the impact is likely to change accordingly. One detailed proposal has been presented to the City Council of Dordrecht in the Netherlands. This experiment would compare the labour market behaviour of a control group staying within the present system and a group of workers and claimants joining a basic income scheme giving them the same net income but permitting them freely to combine their basic income with any amount of paid and unpaid activities. However, at the time of writing, this is still in its planning stages with no firm commitment to proceed. There is an urgent need, therefore, for such pilot projects to be conducted.

Even if such pilot projects are undertaken, however, it is necessary to recognise that basic income schemes go against the grain of current discourse on the future of work and welfare so far as governments in the advanced economies are concerned. Given this, it might be the case that greater attention should be paid to the grass-roots side of the argument which seeks to enable people to help themselves and is very much in line with current discourse in the advanced economies. As Lipietz (1992: 95) puts it, 'In essence, a universal basic allowance is a utopia; this should be understood not just as a desirable state of affairs, but . . . as a compass, a pointer to the path to take'. Here, therefore, we evaluate critically whether the promotion of these grass-roots initiatives would help eradicate informal employment in the advanced economies and restructure work and welfare in the ways suggested by the new economics.

The principal advantage of this self-reliance approach is that it would in theory give people greater ability to meet their needs and wants without having to turn to informal employment. As we have seen in the case of LETS, this grass-roots initiative not only provides the economically excluded with free access to credit but also meaningful and productive work and stronger social networks that they can call upon for help. In practice, however, there is no guarantee that current patterns of social exclusion would disappear. A principal problem with this grass-roots approach is that there is a danger that the 'cart is being put before the horse'. That is, there is a wide range of self-help initiatives being suggested which are intended to provide the means to help people help themselves, but these solutions on the whole are based on a poverty of structured research concerning the barriers to engaging in self-help. There is thus perhaps a need for greater investigation of the nature and extent of such barriers before this approach races ahead with its proposals. At the time of writing, one exception is a piece of research about to be conducted by the authors for the Joseph Rowntree Foundation under its Area Regeneration programme on precisely this issue. This will investigate the barriers to engagement in informal economic activity amongst various economically-excluded groups and how these vary in a range of localities as well as attempting to identify what these populations feel is necessary and desirable to help them help themselves.[2]

Even if these barriers are identified, nevertheless, the issue which remains is again one of whether there is the 'political will' to provide the assistance required to enable people to help themselves. Superficially, it may seem that this is not the case. However, given the difficulties being encountered by governments throughout the advanced economies in financing their welfare states and providing productive and meaningful work for citizens, there is at present much interest in finding creative and innovative solutions to the work and welfare crisis. Much of the current focus from supranational bodies such as the European Commission as well as national and local governments has started to turn towards the potential of harnessing informal economic activity.[3] Nevertheless, despite self-help receiving widespread support across the political spectrum, no locality, region or nation exists which has acted as the site for an experiment in pursuing this type of future and thus, there are no examples for other places to follow. The result is that most local, regional and national governments, as well as most academic commentators, feel that this is either an unrealistic route to follow or at best, a partial solution to the present crises of work and welfare.

Based on this recognition, one organisation has decided to react to this situation by providing an exemplar for others to follow. Forum for the Future (FFF) has chosen West Yorkshire in the UK as the 'laboratory' for its 'Reinvigorating the Local Economy' project which attempts to develop on a larger scale than has so far been witnessed in any area the full raft of grass-roots initiatives in the fields of energy, credit, money, agriculture and land, housing, and business and consumer purchasing. Commencing in 1997 and based in the Royal Society for the Arts northern headquarters in Halifax, the intention is to use West York-shire as the crucible for theorising and implementing such a new economics vision of the future of work and welfare. To date, the 'baseline study' has been completed which has identified the current range of self-help projects taking place in this sub-region and how these relate to the current public sector policy in this area (Williams 1997, 1998). Broad agreement has also been reached with many of the major agencies in the sub-region to pursue this vision and staff have been appointed to start to identify new projects which could be implemented and how currently existing projects could be further developed so as to help provide greater self-reliance in this region. The outcome of this pro-ject, however, and whether it will provide an example for others to follow, will only be known in several years' time.[4] For the moment, therefore, the jury is still out on the feasibility of self-help as a principal solution to the work and welfare crises confronting the advanced economies.

Conclusions

This chapter has presented the third and final approach towards informal employment in the advanced economies. This is the new economics approach which, like the deregulatory approach, views the existence of informal employ-ment as an indictment of the present system, but not of the overregulation of the economy and welfare provision by the state. Instead, it is an indictment of the present system of distributing income and work in society, of which exploitative informal employment is a symptom. At the heart of this approach is the belief that the age of full employment and the comprehensive welfare state, if it ever existed, has gone, and that we need to be more creative in our thinking about the future of work and welfare. Consequently, this approach advocates a radical restructuring of both work and income and, in so doing, seeks to abolish the necessity for people to undertake exploitative unregulated informal employment whilst liberating them to participate more fully in other forms of work.

To achieve this, it advocates both the introduction of a basic income scheme and greater assistance to help people help themselves. This, it believes, will result in the active promotion not only of formal employment but also unpaid informal work. This approach, nevertheless, in contrast to the deregulationist advocacy of self-help, is based on the view that such unpaid informal work can be used to supplement employment and welfare provision, not substitute for it, and that people should have a choice over what form of work they use to meet a need or want rather than an obligation or duty. As such, informal work is seen as a complementary means of meeting needs and wants, not an alternative.

Neither does it advocate the expansion of informal employment. Rather, it assumes that by restructuring work and income in a way which gives people greater choice over whether to use formal employment or unpaid informal work, the necessity for people to engage in exploitative forms of informal employment will be negated.

As we have shown, however, there are major difficulties over the practicality and desirability of implementing this approach to informal employment. So far as the basic income scheme is concerned, there are many imponderables concerning the cost of implementing it, whether there is the political will to do so and whether it will significantly reduce either the level of informal employment in general or exploitative informal employment more particularly. So far as the grass-roots initiatives are concerned, moreover, there is the issue that the new economists are racing ahead with implementing ways of helping people to help themselves before they have a clear conception of the current barriers to the participation of different groups in self-help. Despite being in keeping with both right- and left-wing thinking on work and welfare, there is currently no exemplar of how such an approach can work in practice. Until this occurs, the vast majority of people will probably remain very cautious about the feasibility of the new economics.

12

CONCLUSIONS: RE-PLACING INFORMAL EMPLOYMENT IN THE ADVANCED ECONOMIES

Introduction

The fact that there is a substantial amount of informal employment taking place in the advanced economies is without question. However, whether it is growing or declining, who is doing it and what should be done about it are subjects of heated debate. These issues lie at the heart of the contemporary discourse on informal employment in academic, media and political circles. As such, they are the key questions that this book has sought to address. Reviewing the vast array of studies of informal employment conducted throughout the advanced economies, we have evaluated critically the emerging orthodoxy that the advanced economies are witnessing the growth of informal employment (the informalisation thesis) as it becomes an increasingly important survival strategy for marginalised populations such as the unemployed, ethnic minorities and immigrants (the marginality thesis), as well as the view that stricter enforcement of stringent laws and regulations concerning benefit fraud, tax evasion and the contravention of labour laws (the regulatory approach) is the most appropriate policy option.

Evaluating this emerging orthodoxy through an analysis of the existing evidence on informal employment, the preliminary issue confronted has had to be the quality and validity of the data so far generated in studies of this phenomenon. Although there is some consensus that informal employment can be defined as the paid production and sale of goods and services that are unregistered by, or hidden from the state for tax, social security and labour law purposes, as shown in chapter 1, there are inherent problems with many of the techniques employed to investigate this phenomenon.

Chapter 2 highlighted the fact that more indirect methods which search for evidence of informal employment in macro-economic data collected and/or constructed for other purposes are limited in their usefulness because of their broad-brush approach to the subject. It also revealed that micro-social methods, although providing rich data, are problematic because their results cannot necessarily be taken as evidence of general trends or principles. Consequently, this book has urged that great attention should be paid to the methods employed before reading off trends from any results. Unless this is done, then all attempts to review the composition and size of informal employment will end in confusion over how to interpret apparently contradictory findings. It is only by unpacking the methods utilised in each study and interpreting their results in terms of the limitations of the methods and the context in which it was undertaken that clarity, and understanding of informal employment, can be achieved.

Here, therefore, in this concluding chapter, we review our findings concerning the growth and character of informal employment in the advanced economies, and the policy options relating to this form of work.

Re-placing the growth and character of informal employment

To review our findings concerning the magnitude, growth and character of informal employment in the advanced economies, this section first, analyses the formalisation/informalisation debate, second, analyses the marginality thesis and third, shows how informal employment in the advanced economies can only be explained by *re-placing* this phenomenon, by which we mean grounding explanation in the geographical contexts in which such employment occurs. Finally, we consider the applicability of our retheorisation of informal employment in the advanced economies to the developing nations.

Formalisation, informalisation or heterogeneous development?

A widely-held belief about economic development, and one which for a long time was seldom questioned, is that as economies become more advanced, there is a natural and inevitable shift of economic activity from the informal to the formal sphere (which we have here called the formalisation thesis). Indeed, it is this formalisation of economic and social life which has often been used as the 'measuring rod' to define Third World nations as 'developing' and the First World as 'advanced'. In this view, the existence of so-called 'traditional' informal activities is seen as a manifestation of backwardness that will disappear as advancement and modernisation occurs. Chapter 3, however, showed that the supposedly natural culmination of formalisation – full employment and a comprehensive welfare state – can no longer be accepted as the end-state of economic development. On the one hand, the advanced economies appear to be in a 'post-formalisation' era, as evidenced by the shift away from full employment towards structural unemployment, underemployment and the concentration of formal jobs in a smaller proportion of households, resulting in tremendous pressure on the welfare 'safety net'. On the other hand, the evidence from the developing nations shows that employment is by no means universally growing. There are whole swathes of the 'developing' world where formal employment is not only at a very low base level of about 5–10 per cent of total employment, but where the number of formal jobs are either static or failing to keep pace with the growth in the workforce (see chapter 8). Therefore, the formalisation thesis, upon which many of the current assumptions about the future of work and welfare are founded, does not appear to be justified.

Consequently, it might be assumed that the opposite is taking place. If the advanced economies are witnessing the exclusion of an increasing proportion of their population from the formal sphere of work and welfare, then they must be turning towards informal employment as a means of survival. This is the informalisation thesis. For its adherents, informal employment thus marks a new and expanding form of advanced capitalism rather than a mere lag from traditional relationships of production. The problem identified in chapter 3 however, is that there is little reliable evidence to support this notion of a universal informalisation of the advanced economies. The indirect macro-economic methods suffer from fatal flaws which make the evidence unreliable whilst the micro-social studies, although more direct and accurate in their portrayals of such work, provide little if any evidence of a general informalisation of the advanced economies. Most are snapshots taken at a particular point in time and those few that do provide longitudinal data supply no evidence of increased informalisation (Fortin et al. 1996, Mogensen et al. 1995). Neither can the deformalisation of the advanced economies be taken as proof of informalisation. To do this is to assume that formal and informal work are substitutes for each other and that the rise of one leads to the decline of the other. The problem, however, is that although this is the case in some places at certain moments in time (Castells and Portes 1989, Duncan 1992, Gutmann 1978), it is not universally the case. Numerous other studies suggest that the relationship is one of complementarity, not substitution, in that formality and informality either grow or decline simultaneously (Cappechi 1989, Cornuel and Duriez 1985, Leonard 1994, Morris 1995, Pahl 1984). The relationship between formality and informality therefore appears to be far more complex than simply one of universal substitutability or even complementarity. The demise of formal work and welfare is thus not everywhere necessarily leading to a relative informalisation of the advanced economies.

Instead, we have introduced the idea that geography matters. Examining the findings of the vast array of micro-social studies, it has been shown that there are different processes in different places at varying times. As such, it is more accurate to describe the development trajectory of the advanced economies as one of heterogeneous development, with different places undergoing either informalisation or formalisation at varying times and at different rates. How we have explained this is returned to below. Before doing this, it is first necessary to evaluate critically the second emerging orthodoxy: that informal employment is a type of peripheral job at the bottom of a hierarchy of flexible types of employment and is undertaken by marginalised populations as a survival strategy (the marginality thesis).

Informal employment: a place for the marginalised or a segmented informal labour market?

Throughout this book, we have shown that far from being a peripheral form of flexible employment solely for the marginalised, a wide range of people engage in informal employment of contrasting types (e.g. organised/autonomous, part-time/full-time, permanent/temporary, skilled/unskilled) for diverse reasons (ranging from solely economic to entirely social) and for substantially different rates of pay. As such, we have retheorised informal employment as a multifarious phenomenon. Existing alongside what everybody accepts as a heterogeneous formal labour market, we have shown that there is a less recognised heterogeneous informal labour market. This informal labour market, to adopt a simplistic dual labour market model, ranges from 'core' informal employment which is relatively well-paid, autonomous and non-routine and where the worker often benefits just as much from the work as the employer, to 'peripheral' informal employment which is poorly-paid, exploitative and routine and where the employee does not benefit as much as the employer.

Using this notion of a segmented informal labour market, we then proceeded to examine the level and character of informal employment by employment status (chapter 4), gender (chapter 5), ethnicity and immigration (chapter 6) and geography (chapter 7). This revealed that throughout the advanced economies the unemployed, ethnic minorities, immigrants, women and people living in poorer localities generally engage in less informal employment than the employed, dominant white population, men and people living in more affluent areas. Furthermore, when they do engage in informal employment, these marginal groups find themselves confined to low-paid, exploitative, peripheral, informal employment which they undertake for primarily economic reasons, whilst the better-paid, autonomous, informal work is mostly undertaken by the employed, dominant white population, men and people living in more affluent areas. Consequently, if marginal groups find themselves in exploitative and low-paid informal employment, it is because they find themselves in this kind of work in formal employment, not because informal employment is comprised of this sort of work alone and these sort of workers only. Put another way, informal employment usually reinforces rather than mitigates the plight of marginal groups.

However, although a segmented informal labour market appears to be applicable in all localities, regions and nations, the precise configuration of this labour market is not. In some areas, not only will the proportion of such work which is core and peripheral vary greatly but so will the precise composition of the core and peripheral informal labour force. Put another way, although the informal labour market is normally structured in the way described above, which reflects the formal labour market, it is not always configured in this manner. Exceptions exist where alternative configurations prevail, such as where the unemployed, women or ethnic minorities conduct more informal employment than the employed, men or dominant white population, or where some sections of these marginal groups achieve social mobility in the form of entry into the core informal labour force. Therefore, not only does the magnitude and growth of informal employment vary from place to place, but so does its character.

Explaining the heterogeneous growth and character of informal employment

To explain why the growth, magnitude and character of informal employment varies from place to place, we have argued that this is a product of how economic, social, institutional and environmental conditions combine in multifarious 'cocktails' in different places to produce specific local outcomes. The consequence is that at any time, there is neither universal formalisation nor informalisation across the advanced economies but, rather, different processes in varying places. Neither is there one structure of the informal labour market which is the same across all places. Instead, different configurations prevail in different areas.

Our intention in arguing for a recognition of the heterogeneous magnitude, development and character of informal employment in different areas, however, has not been to compensate for the previously overgeneralised and simplistic typologies concerning the magnitude and character of informal employment by concluding with an idiographic approach which descends into particularism. This would be to go too far to the other extreme, ignoring some important parallels and commonalities in the nature and extent of informal employment between localities, regions and nations both in the advanced economies and beyond. Instead, our aim has been to draw out the commonalities and explain them, whilst attempting not to lose sight of the variations, both in terms of the level and growth of informal employment in different areas (see chapter 7) and in how this informal labour market is often similarly configured (see chapters 4–6).

In the chapters on socio-spatial divisions of labour in Part II, we have thus shown not only how the most popular 'cocktails' to be found in the advanced economies produce a segmented informal labour market which reinforces the inequalities produced by the formal labour market but also how there are particular situations in which the 'cocktails' combine together in different ways to produce alternative configurations of informal employment. As such, we call for a re-placing of informal employment. That is, we argue that there are no universal tendencies so far as either the level or character of informal employment are concerned but, rather, different processes in different areas. To understand informal employment, therefore, the need is to explore the processes by which economic, social, institutional and environmental factors variously combine in different areas to produce particular configurations of informal employment. Without such a geographically-refined approach to informal employment, the previous mistake of seeking over-simplistic and universal tendencies will continue to plague the subject.

Applications to informal employment beyond the advanced economies

Having outlined this need for a re-placing of informal employment in the advanced economies so that both its heterogeneity and the commonalities between areas can be recognised, a particular concern of this book has been to ask whether such a re-placing is also required in the developing nations. Are the developing economies witnessing formalisation, as modernisation theory would have us assume, or is there a process of informalisation taking place? Alternatively, is such an either/or choice too simplistic to capture the diverse experiences of this large portion of the world? Moreover, is the 'marginality thesis' applicable to informal employment across the entire developing world? Is such employment merely an exploitative form of peripheral work existing amongst marginalised groups or is there a heterogeneous informal labour market with a hierarchy of its own? If so, does it mitigate or reinforce the socio-spatial divisions prevalent in the formal labour market?

Addressing these questions through a review of the literature on informal employment in the developing nations, chapter 8 has shown that it is far too simple to assert that there is either a formalisation or informalisation across all localities, regions and nations in the developing nations. Rather, we have shown that there is a process of heterogeneous development whereby at any time, some areas are formalising their economies whilst others are becoming more informalised. Moreover, there also appear to be similarities between the configuration of informal employment in the developing and advanced economies. That is, just as in the advanced economies one cannot simply view informal employment as a refuge for marginalised populations (the marginality thesis), the same applies in developing nations. Instead, there appears to be a segmented informal labour market with a hierarchy of its own. This hierarchy, moreover, seems to reproduce, rather than mitigate, the socio-spatial inequalities prevalent in the formal labour market.

Given that informal employment is at least as significant as formal employment, if not more so, for the majority of the population in these nations and that the developing economies constitute the vast majority of the world, such a rethinking of informal employment is of crucial importance. It is essential not only for understanding a major segment of the global economy but also for identifying what can be done to improve the well-being of the majority of the citizens of the world. Inasmuch as the central focus of this book is the advanced economies, such policy outputs have not been specified for the developing nations. Instead, our focus has been upon the advanced economies. Neverthe-less, those involved in the study of developing nations can judge for themselves whether the policy options we have discussed in Part III in relation to the advanced economies would have similar impacts if implemented in the develop-ing nations.

Re-placing informal employment in the advanced economies: the implications of alternative policy options for work and welfare

Having shown in Part II how informal employment generally tends to reinforce rather than reduce the socio-spatial disparities produced by formal employ-ment, Part III of this book has evaluated critically the various options open to policy-makers. It is important to be clear that few, if any, commentators advo-cate the continuation or development of informal employment as a whole in its present form. All commentators wish to replace informal employment. How-ever, each approach seeks to replace it with different forms of work and in different ways, depending upon its vision for the future of work and welfare.

Examining the approaches towards informal employment advocated in the literature, Part III has thus discerned three principal policy options. The first two approaches, of either regulating informal employment or deregulating formal employment, both seek to replace informal with formal employment, whilst the third, the new economics, seeks to replace such employment with a combination of unpaid informal work and formal jobs. In Part III, the impli-cations for work and welfare of pursuing each of these approaches have been evaluated critically in turn. In each case, first, we have described in detail the main arguments put forward within the approach concerning informal employ-ment itself, and the ways in which these arguments are grounded in a particular vision of the future of work and welfare. Subsequently, we have critically evaluated these arguments from the point of view of their practicability and desirability.

Starting with the regulatory approach, this seeks to replace informal employ-ment with formal employment and the method it adopts to achieve this is tougher enforcement of labour law, welfare benefits and tax rules. The hope is that in so doing, informal jobs will be turned into formal jobs and that full employment will return. Superficially, it may seem to many readers that this is the most appropriate policy response. After all, as shown in Part II, informal employment generally reinforces rather than reduces the socio-spatial divisions produced by formal employment. However, such an approach has been shown to be not as unproblematic as it may initially appear. Despite all of the evidence to the contrary, this approach retains a blind faith in a return to full employ-ment and comprehensive welfare provision. Little consideration is given to the notion that this might not be achievable and thus that people need to be given alternative ways of engaging in meaningful and productive activity both as a social entitlement and as part of a policy of active citizenship. Examining the feasibility of eradicating informal employment, moreover, it has been shown to be impractical given the barriers which confront policy-makers, and is also undesirable due to the fact that it would have the unintended effect of widening socio-spatial disparities.

An alternative to the stricter enforcement of laws and regulations is to attempt to apply other forms of proactive pubic policy that might enable informal employment to be turned into formal jobs. The deregulatory approach, similar to the regulatory approach, again desires a return to 'full employment' but rather than bolstering the flagging edifices of 'diluted' or welfare capitalism through state-led work-sharing and job creation as well as a comprehensive welfare state, instead advocates 'undiluted' capitalism so as to 'let the market rip' and roll back the welfare state. The intention, in so doing, is to close the gap between formal and informal employment by eradicating the rules and regulations that hinder the achievement of full employment. As such, the already deregulated sphere of informal employment is seen as an exemplar of how work and welfare should be organised. The aim however, is not to promote informal employment as it currently exists. Instead, it seeks to replace the formal/informal dichotomy not by regulating informal employment out of existence but by deregulating the formal side of the equation so as to abolish the distinction. Although there is little doubt that deregulation of formal employment would reduce the magnitude of informal employment, it is very doubtful that deregulation of work and welfare would enable the socially excluded to improve their circumstances. Instead, the evidence throughout the advanced economies is that social inequalities widen at a quicker rate in nations pursuing deregulation and that deregulation does little to help marginalised groups and areas to solve their problems. In sum, although deregulation is likely to lead to the eradication of informal employment in the advanced economies, the overall impact is a levelling down rather than up of material and social circumstances.

The final approach towards informal employment is that of the new economics. Unlike the previous two approaches, this does not seek to replace informal employment with formal jobs but, rather, with a combination of unpaid work and formal employment. Informal employment is here seen as an indictment of the present system, but not of the overregulation of the economy and welfare provision by the state. Instead, it is an indictment of the present system of distributing income and work in society, of which exploitative informal employment is a symptom. Based on the view that full employment will not return, this approach advocates a radical restructuring of both work and income and, in so doing, seeks to abolish the necessity for people to undertake exploitative informal employment whilst liberating them to participate more fully in other forms of work.

To achieve this, it advocates the introduction of a basic income scheme and greater assistance to help people help themselves. This, it believes, will result in the active promotion not only of formal employment but also of unpaid informal work. Unlike the deregulationist advocacy of self-help, however, this approach is based on the view that such informal work can be used to supplement employment and welfare provision, not substitute for it, and that people should have a choice over what form of work they use to meet a need or want rather than an obligation or duty. As such, informal work is seen as a complementary means of meeting needs and wants, not an alternative. Neither does it advocate the expansion of informal employment. Rather, it assumes that by restructuring work and income in a way which gives people greater choice over whether to use formal employment or unpaid informal work, the necessity for people to engage in exploitative forms of informal employment will be negated.

As shown in chapter 11, however, there are immense difficulties over the practicality and desirability of implementing this approach to informal employment. So far as the basic income scheme is concerned, there are many imponderables concerning not only both the cost of implementing it and whether there is the political will to do so, but also whether it will significantly reduce either the level of informal employment in general or exploitative informal employment more particularly. So far as the grass-roots initiatives are concerned, moreover, there is the issue that the new economists are racing ahead with implementing ways of helping people to help themselves before they have a clear conception of the current barriers to the participation of different groups in self-help. Neither is there currently an exemplar of how such an approach can work in practice, nor even a crucible where such experiments are being tested on a large scale. The implication is that despite being in keeping with much of the new thinking on work and welfare from both left- and right-wing commentators, there remain major reservations about the feasibility of such an approach.

In sum, and having reviewed three different approaches towards informal employment in the advanced economies, we have shown that the choice of which to adopt is very much dependent upon the wider issue of the future for work and welfare we would like. Do we want try to return to the commodification of both work and welfare in the form of full employment and the welfare state, despite all of the evidence suggesting that this is not possible? Or, alternatively, do we want a future of work and welfare in which the formal labour market is deregulated so as to reduce wage rates, to return to full employment whilst taking away the 'safety net' to give people the clear message that they must pull themselves up by their own bootstraps? Or should we accept that full employment is gone and seek to restructure work and welfare using the whole economy and do so in manner which helps people to help themselves and pursue greater self-reliance? These are ultimately the questions that will need to be answered before one can decide which is the appropriate policy towards informal employment in the advanced economies. It is certain, however, that a *laissez-faire* policy is not an option. To leave it alone will do little except reinforce the social and spatial inequalities and injustices already being perpetrated.

NOTES

1 Introduction

1 For further criticisms of using nouns such as 'economy' and 'sector' to define informal employment, see Harding and Jenkins (1989) and Roldan (1985).
2 By examining solely goods and services, this definition of informal employment also excludes purely financial fraud.
3 One indication of the growing interest in informal employment is that during 1997, the major government funding vehicle for social science research in the UK, the Economic and Social Research Council (ESRC), set in motion a £1.8 million Research Programme on Informal Economic Activity which has at its heart a focus upon both tax evasion and social security fraud. Further examples of the growing interest by the state in informal employment is that in late 1997, the European Commission produced a Green Paper on Illegal Work (not published at the time of completion of this manuscript) and that by October 1998, each nation in the European Union (EU) is expected to provide an estimate of the adjustments required to GNP due to the existence of informal employment in their country.
4 One exception is the paper by Windebank and Williams (1997).

2 Methods of researching informal employment

1 See Thomas (1988, 1992) for a more detailed critique of the indirect monetary methods than is possible here, whilst a more in-depth critique of the income/expenditure discrepancies method can be found in Smith (1986).
2 These, however, have not been without problems. In Portugal, for example, Santos employs Gutmann's method to estimate the level of informal employment which he puts at 22 per cent of GDP in 1981 in one report (Santos 1983) but 11.2 per cent for the same year in another (Cocco and Santos 1984). If the same author comes to such remarkably different conclusions in two reports, then grave doubts must exist over the accuracy of this method.
3 See chapter 4 for a critique of the assumption that there is a causal relationship between rises in unemployment levels and the volume of informal employment.

3 Theorising informal employment

1 Indeed, our whole notion of 'family' and its role in the advanced economies depends upon continued informality. That is, the family has nowhere surrendered anywhere near all of its welfare and economic functions to the formal sphere.
2 One indicator of the EU's concern over whether formal welfare provision can expand or even be maintained is that it is currently funding two major projects under its Targeted Socio-Economic Research (TSER) Programme to explore the viability of utilising informal economic activity as a means of both combating social exclusion and promoting social inclusion. For further details on these projects, that run until 2000, contact the authors.
3 This is not to say that such micro-social studies would not reveal such a process of informalisation if more longitudinal studies were conducted. In this regard, one particularly useful method is the time-budget diaries technique, which could explore the changing participation in informal employment both absolutely and relative to other forms of work (see Gershuny *et al.* 1994). Unfortunately, however, despite a number of longitudinal studies examining the paid/unpaid work balance, no such studies have so far examined informal employment. We would like to thank Jonathan Gershuny for his personal correspondence confirming that this is indeed the case.

4 Employment status and informal employment

1 As Cook (1997) shows in the UK, the result is a two-tier legal system in which there is one law for the rich and one for the poor. A blind eye is turned to multi-million pound 'white-collar' fraud whilst there is an increasing concentration on catching the poor. She states that in the year ending April 1995, despite £6 billion being recovered in unpaid taxes, only 357 people were prosecuted, compared with 9,546 fraud cases mounted by the DSS with only £650 million saved. Most of the latter cases involved small amounts of money. As the OECD (1994) reports, moreover, this two-tier approach is not unique to the UK. It has been a common trend throughout many advanced economies.

5 Gender and informal employment

1 An exception is Hoyman (1987).
2 Here, however, it needs to be recognised that skill levels and hierarchies are socially constructed (Crompton *et al.* 1996, Rubery 1996). How a skill is defined is influenced by employer beliefs, patriarchal ideologies and the capacity of organisations representing employees to shape skill classifications, to name but a few influences. Indeed, much of what is conventionally seen to be unskilled informal work may indeed be skilled, but because it is conducted by women or marginal groups is deemed not to be so.
3 It is here recognised that although we treat the domestic mode of production as a 'social' rather than 'economic' characteristic, for some feminists, it is itself viewed as an 'economic' mode of production characterised by patriarchy which operates alongside the formal mode of production characterised by capitalism (e.g. Delphy 1984).
4. This notion of 'lagged adaptation' is further reinforced by examining the husband's proportion of the household's total work by the length of time that the wife has been employed over the last ten years. Gershuny *et al.* (1994) find that

183

in households where the wife has twenty months or less employment experience, the total of the couple's work is divided in the ratio 46:54 between the husband and wife; over the range twenty-one to sixty months in employment the total work is divided 48:52; and where the wife has more than five years work experience, the total work is evenly divided.

5 Unfortunately, there is currently no systematically collected comparable empirical data either on cross-national or intranational spatial variations in the configuration of informal employment to allow such hypotheses to be tested. For the moment, therefore, they must remain simply assertions about the likely spatial variations in the gender configuration of informal employment.

6 Ethnicity, immigration and informal employment

1. Studies comparing the labour market achievement of central city to suburban African-Americans, moreover, suggest that the economic gap between them is widening over time (Jencks and Mayer 1990, Kasarda 1989), even though blacks who do suburbanise tend to cluster in segregated pockets in the older inner suburbs rather then dispersing throughout the newer predominantly white outer suburbs (Galster 1991, Schneider and Phelan 1993).

2 An Economic and Social Research Council (ESRC) funded project currently underway is attempting to answer many of the questions posed here. Examining undocumented Brazilian workers in London and Berlin, the pilot study revealed that many were well educated but had limited or poor prospects in Brazil. Their accounts of their experiences and decisions mixed personal development (they entered as tourists) with investment (education, including language and work skills) and earning (often to support the first two). Investigating their strategies for staying and working, it was identified that Brazilians in London integrated through informal labour and housing markets (but hardly at all culturally) whilst Brazilians in Berlin relied on relationships with Germans and legal/settled immigrants, and could not get work or accommodation without these (Jordan *et al.* 1992). The present ESRC-funded study is examining the impact of immigration control policies and practices as well as the motives and experiences of Turkish and East European immigrants, and should go some way towards answering many of the questions we have posed.

7 Spatial divisions in informal employment

1 It is necessary to speak of the 'underground economy' when discussing these macro-economic studies because they cannot, on the whole, distinguish between criminal activities and informal employment. The term 'underground economy' is most commonly used to refer to both these categories as a unit.

2 At the time of writing, the European Commission's Targeted Socio-Economic Research (TSER) programme has agreed to fund a project which will conduct a standardised questionnaire of informal economic activity across six EU countries. Entitled INPART (Inclusion through Participation), further information on this project, co-ordinated by the University of Utrecht, can be obtained from the authors who are conducting the UK element of the study. The full cross-national findings are likely to be available in early 2000. A further study of the volume and nature of informal employment in different kinds of type 3 area is currently being conducted in the UK under the auspices of a Joseph Rowntree Foundation grant as part of its Area Regeneration programme and a further study, again of a

type 3 area, looks set to be conducted by the Comprehensive Estates Initiative of Hackney Borough Council in London. At present, nevertheless, there is still no study which compares, in a systematic manner, using the same survey method, the magnitude and character of informal employment in each of these area-types. This is urgently required to test the validity of this typology so that further progress can be made on mapping out a geography of informal employment.

8 Informal employment in developing nations

1 A similar complementary relationship between formal and informal employment has been identified in Montevideo (Uruguay), where Grosskoff and Melgar (1990) find that the growth of one leads to the growth of the other, and vice versa.
2 As Cheng and Gereffi (1994: 204) assert, 'it is only by nesting the informal economy in the unique developmental context of . . . [each nation] that we can appreciate its nature and significance'.
3 Indeed, there are many regions of the world where formalisation has hardly even taken hold. In low-income nations, for example, regular waged or salaried employment 'accounts for only 5 to 10 per cent of total employment' (International Labour Organisation 1996: 140).
4 It should be noted, nevertheless, that although Hart (1973) is usually accredited with developing the notion of a formal/informal dichotomy, he was not the first person to distinguish a dual economy. Indeed, this dual economy concept has been circulating for some forty years (e.g. Lewis 1959, Polanyi 1957).
5 In other regions, however, land ownership patterns influence the gendering of migration in different ways. As Thomas (1992) asserts, in some parts of Asia (e.g. the Philippines) and in Latin America, many families do not own or have claim to any specific land and if the demand for rural labour is mainly for men, the opportunity costs of migration may be lower for women than for men. Therefore, the gendering of the informal labour market cannot be explained by women's responsibility for unpaid domestic work alone. Instead, other factors such as land ownership systems mediate the way in which the gender divisions of informal employment are manifested in practice. Nevertheless, if women's participation rates are influenced by many other factors besides women's responsibility for unpaid domestic work, the nature of the informal employment that they undertake remains heavily shaped by this factor. Women everywhere in the developing nations conduct informal employment that fits their domestic roles and reflects their domestic duties and responsibilities.

9 Regulating informal employment

1 In this sense, these analysts who view exploitative informal employment as a 'revenge of the market' are similar to the deregulationists discussed in the next chapter. However, whilst these analysts view this as a negative development, the deregulationists hold up such work up as an exemplar of the way forwards.
2 The result is that in 1990, employment accounted for an average of 70,000 hours out of a lifetime of 640,000 hours, that is, 11 per cent of total time for the EU population. In a few decades from now, the EU estimates that the proportion of one's lifetime spent employed will drop to some 40,000 hours, i.e. 6 per cent of a total lifespan of 700,000 hours (European Commission 1994).

3 This household-based approach to examining social polarisation in employment opportunities has been most prevalent in the UK (Dex and Taylor 1994, Dorling and Woodward 1996, Jarvis 1997, Pahl 1984, Pinch 1994, Williams and Windebank 1995b, Woodward 1995). Here, there has been a tendency to view such polarisation as being an effect of social security policy. The argument has been that wives of unemployed men cannot take part-time jobs (the major growth sector) due to the effect on the man's entitlement to benefits and that this is leading to a polarisation of households (see Hewitt 1996, Morris 1995). However, the fact that such polarisation is taking place in many other OECD nations casts into doubt such nationally-orientated institutionalist explanations. It may thus be the case that polarisation relates much more to the ways in which households are composed of people with comparable human capital assets and that those with less marketable skills are suffering across OECD nations from the universal economic restructuring processes taking place (Davies et al. 1992).

4 Further reasons why all informal employment will not convert into formal jobs under tougher regulatory conditions are given by Peacock and Shaw (1982). They argue that to calculate the tax loss by multiplying the size of informal employment by some average effective tax rate produces an overestimate for three reasons. First, some activities that are profitable in informal employment at the 'evasion market wage' would not be viable if tax had to be paid. Since these activities would disappear if tax were charged, they should not be included in the tax loss calculations. Second, such a calculation ignores the fact that some of the income of tax evaders is spent on formal goods and services and is thus subject to indirect taxation, which generates revenue for the government. Finally, there is the fact that expenditure by tax-evaders generates incomes for others, including honest tax-payers. Revenues are then generated when these individuals pay their taxes.

10 Deregulating formal employment

1 One has only to look at how the 'Chicago boys' implemented such a neo-liberal ideology in Chile to see that this academic theorisation has major implications for whole societies. As Huber (1996) reveals, Chile's poverty rate rose from 17 per cent in 1970 to 38 per cent in 1986, with the unemployment rate in 1983 reaching a third of the labour force.

11 Informal employment and the new economics

1 The New Economics Foundation can be contacted at: 1st Floor, Vine Court, 112–16 Whitechapel Road, London E1 1JE.

2 For further details of this project, entitled 'Developing the informal sector in deprived neighbourhoods: towards a complementary approach to local economic development', funded by the Joseph Rowntree Foundation under their Area Regeneration programme, contact the authors.

3 For example, the European Commission under its Targeted Socio-Economic Research (TSER) programme is currently funding two initiatives to explore whether informal economic activity can provide a tool for enabling social inclusion and tackling social exclusion. The first, co-ordinated by the University of Utrecht and entitled Inclusion through Participation (INPART), aims to evaluate the inclusion potentials of both paid and unpaid work for different social groups. The second project, co-ordinated by the Political Economy Research Centre

(PERC) at the University of Sheffield, entitled 'Comparative social inclusion policy and citizenship in Europe: towards a new European social model', again evaluates the potential use of forms of work other than employment in the construction of a new social model based on active citizenship. There is also widespread support by local government for a range of self-help initiatives under their anti-poverty programmes, economic development strategies and Local Agenda 21 policies. To take just one example, some 60 per cent of local authorities in the UK have an explicit statement of support for LETS (see Williams 1996d).

4 Further information on this 'Reinvigorating the local economy' project of Forum for the Future is available either from Colin Williams who has conducted the baseline study and is a non-executive director of this project, or direct from Forum for the Future, 227a City Road, London EC1V 1JT.

REFERENCES

Abrahamson, P. (1992) 'Welfare pluralism: towards a new consensus for a European social policy?', in L. Hantrais, M. O'Brien and S. Mangen (eds) *The Mixed Economy of Welfare*, Cross-National Research Paper no. 6, Loughborough: European Research Centre, Loughborough University.

Addison, T. and Demery, L. (1987) *The Alleviation of Poverty under Structural Adjustment*, Washington, DC: World Bank.

Aitken, S. and Bonneville, E. (1980) *A General Taxpayer Opinion Survey*, Washington, DC: CSR Inc.

Alden, J. (1982) 'A comparative analysis of moonlighting in Great Britain and the USA', *Industrial Relations Journal* 13: 21–31.

Amado, J. and Stoffaes, C. (1980) 'Vers une socio-economie duale', in A. Danzin, A. Boublil and J. Lagarde (eds) *La Société Française et la technologie*, Paris: Documentation Française.

Amin, A. (1994) 'The difficult transition from informal economy to Marshallian industrial district', *Area* 26,1: 13–24.

Amin, S. (1996) 'On development: for Gunder Frank', in S.C. Chew and R.A. Denemark (eds) *The Underdevelopment of Development*, London: Sage.

Amott, T. (1992) *Women and the US Economy Today*, New York: Basic Books.

Amott, T. and Matthaei, J. (1991) *Race, Gender and Work: A Multi-cultural Economic History of Women in the United States*, Montreal: Black Rose Books.

Anheier, H.K. (1992) 'Economic environments and differentiation: a comparative study of informal sector economies in Nigeria', *World Development* 20,11: 1573–85.

Atkinson, A. and Micklewright, J. (1991) 'Unemployment compensation and labour market transitions: a critical review', *Journal of Economic Literature* 29: 1679–727.

Atkinson, A.B. (1995) *Public Economics in Action: The Basic Income/Flat Tax Proposal*, Oxford: Oxford University Press.

Aznar, G. (1981) *Tous à mi-temps, ou le Scenario Bleu*, Paris: Seuil.

Barnes, H., North, P. and Walker, P. (1996) *LETS on Low Income*, London: New Economics Foundation.

Barr, N. (1992) 'Economic theory and the welfare state: a survey and interpretation', *Journal of Economic Literature* 30: 741–803.

Barthe, M.A. (1985) 'Chomage, travail au noir et entraide familial', *Consommation* 3: 23–42.

—— (1988) *L'Economie cachée*, Paris: Syros Alternatives.

Barthelemy, P. (1990) 'Le travail au noir en Belgique et au Luxembourg', in European Commission, *Underground Economy and Irregular Forms of Employment*, Brussels: Office for Official Publications of the European Communities.

—— (1991) 'La croissance de l'economie souterraine dans les pays occidentaux: un essai d'interpretation', in J-L. Lespes (ed.) *Les Pratiques juridiques, economiques et sociales informelles*, Paris: PUF.

Baud, I.S.A. (1993) *Forms of Production and Women's Labour: Gender Aspects of Industrialisation in India and Mexico*, London: Sage.

Baxter, J. (1992) 'Domestic labour and income inequality', *Work, Employment and Society* 6,2: 229–49.

Beatson, M. (1995) *Memories of Class*, London: Routledge.

Beechey, V. and Perkins, T. (1987) *A Matter of Hours: Women, Part-time Work and the Labour Market*, Cambridge: Polity.

Bell, C. and McKee, L. (1985) 'Marital and family relations in times of male unemployment', in B. Roberts, R. Finnegan and D. Gallie (eds) *New Approaches to Economic Life*, Manchester: Manchester University Press.

Beneria, L. (1992) 'The Mexican debt crisis: restructuring in the economy and household', in L. Beneria and S. Feldman (eds) *Unequal Burden: Economic Crises, Persistent Poverty, and Women's Work*, Boulder, CO: Westview.

Beneria, L. and Roldan, M.I. (1987) *The Crossroads of Class and Gender: Homework, Subcontracting and Household Dynamics in Mexico City*, Chicago: Chicago University Press.

Beneria, L. and Stimpson, C.R. (1987) (eds) *Women, Households and the Economy*, New Brunswick, NJ: Rutgers University Press.

Bennington, J., Baine, S. and Russell, J. (1992) 'The impact of the Single European Market on regional and local economic development and the voluntary and community sectors', in L. Hantrais, M. O'Brien and S. Mangen (eds) *The Mixed Economy of Welfare*, Cross-National Research Paper no. 6, Loughborough: European Research Centre, Loughborough University.

Benton, L. (1990) *Invisible Factories: The Informal Economy and Industrial Development in Spain*, New York: State University of New York Press.

Berthelier, P. (1993) *Conditions d'activité de la population de Yaounde*, Paris: DIAL-Paris et DSCN-Yaounde.

Bhavnani, R. (1994) *Black Women in the Labour Market: Research Review*, Manchester: Equal Opportunities Commission.

Biggs, T., Grindle, M.D. and Snodgrass, D.R. (1988) *The Informal Sector, Policy Reform and Structural Transformation*, EEPA discussion Paper no. 14, Cambridge, MA: Harvard Institute for International Development.

Blair, J.P. and Endres, C.R. (1994) 'Hidden economic development assets', *Economic Development Quarterly* 8,3: 286–91.

Bloeme, L. and Van Geuns, R.C. (1987) *Ongeregeld Ondernemen: een onderzoek naar informele bedrijvigheid*, The Hague: Ministerie van Sociale Zaken en Werkgelegenheid.

Boris, E. (1996) 'Sexual divisions, gender and constructions: the historical meaning of homework in Western Europe', in E. Boris and E. Prugl (eds) *Homeworkers in Global Perspective: Invisible no More*, London: Routledge.

—— and Prugl, E. (1996) 'Introduction', in E. Boris and E. Prugl (eds) *Homeworkers in Global Perspective: Invisible no More*, London: Routledge.

Borocz, J. (1989) 'Mapping the class structures of state socialism in East-Central Europe', *Research in Social Stratification and Mobility* 8: 279–309.

Brandt, B. (1995) *Whole Life Economics: Revaluing Daily Life*, Philadelphia, PA: New Society Publishers.

Brannen, J., Meszaros, G., Moss, P. and Poland, G. (1994) *Employment and Family Life: A Review of Research in the UK (1980–1994)*, London: Employment Department Research Series no. 41.

Brannen, J. and Moss, P. (1991) *Managing Mothers: Dual Earner Households after Maternity Leave*, London: Macmillan.

Briar, C. (1997) *Working for Women? Gendered Work and Welfare Policies in Twentieth Century Britain*, London: UCL Press.

Briggs, V. (1984) 'Methods of analysis of illegal immigration into the United States', *International Migration Review* 18: 623–41.

Brindle, D. (1995) 'Lilley targets £1.4 bn benefit fraud', *Guardian* 11 July: 5.

Bromley, R. and Gerry, C. (1979) *Casual Work and Poverty in Third World Cities*, Chichester: Wiley.

Brusco, S. (1986) 'Small firms and industrial districts: the experience of Italy', in D. Keeble and F. Waever (eds) *New Firms and Regional Development*, London: Croom Helm.

Bryson, A. and Jacobs, J. (1992) *Policing the Workshy*, Aldershot: Avebury.

Bunker, N. and Dewberry, C. (1984) 'Unemployment behind closed doors', *Journal of Community Education* 2,4: 31–3.

Burawoy, M. and Lukacs, J. (1985) 'Mythologies of work: a comparison of firms in state socialism and advanced capitalism', *American Sociological Review* 50: 723–37.

Button, K. (1984) 'Regional variations in the irregular economy: a study of possible trends', *Regional Studies* 18: 385–92.

Cappechi, V. (1989) 'The informal economy and the development of flexible specialisation in Emilia Romagna', in A. Portes, M. Castells and L.A. Benton (eds) *The Informal Economy: Studies in Advanced and Less Developing Countries*, Baltimore: Johns Hopkins University Press.

Carbonetto, D., Hoyle, J. and Tueros, M. (1987) *El sector informal urbano en Lima Metropolitana*, Lima: CEDEP.

Castells, M. and Portes, A. (1989) 'World underneath: the origins, dynamics and affects of the informal economy', in A. Portes, M. Castells and L.A. Benton (eds) *The Informal Economy: Studies in Advanced and Less Developing Countries*, Baltimore: Johns Hopkins University Press.

Chavdarova, T. (1995) 'Tax morality and households' strategies of economic activity: challenges for social policy in Bulgaria', Paper presented at *Euroconference in Social Policy in an Environment of Insecurity: Contemporary Dilemmas and Challenges for Social Policy*, Lisbon, November.

Cheng, L.L. and Hsuing, P.C. (1991) 'Women, export-oriented growth and the state: the case of Taiwan', in R. Appelbaum and J. Henderson (eds) *States and Development in the Asian Pacific Rim*, Beverley Hills: Sage.

Cheng, L.L. and Gereffi, G. (1994) 'The informal economy in East Asian development', *International Journal of Urban and Regional Research* 18,2: 194–219.

Chu, Y.W. (1992) 'Informal work in Hong Kong', *International Journal of Urban and Regional Research* 16,3: 420–31.

Citizen's Income Trust (1995) *Paying for Citizen's Income*, Pamphlet no. 3, London: Citizen's Income Trust.

Cocco, M.R. and Santos, E. (1984) 'A economia subterranea: contributos para a sua analisee quanticacao no caso Portugues', *Buletin Trimestral do Banco de Portugal* 6,1: 5–15.

Cochrane, A. and Clark, J. (1993) *Comparing Welfare States: Britain in International Context*, London: Sage.

Coffield, F., Borrill, C. and Marshall, S. (1983) 'How young people try to survive being unemployed', *New Society* 2 June: 332–4.

Connolly, P. (1985) 'The politics of the informal sector: a critique', in N. Redclift and E. Mingione (eds) *Beyond Employment: Household, Gender and Subsistence*, Oxford: Basil Blackwell.

Conroy, P. (1996) *Equal Opportunities for All*, European Social Policy Forum Working Paper I, Brussels: European Commission, DG V.

Contini, B. (1982) 'The second economy in Italy', in V.V. Tanzi (ed.) *The Underground Economy in the United States and Abroad*, Lexington, MA: Lexington Books.

Cook, D. (1989) *Rich Law, Poor Law: Different Responses to Tax and Supplementary Benefit Fraud*, Milton Keynes: Open University Press.

—— (1997) *Poverty, Crime and Punishment*, London: Child Poverty Action Group.

Cornuel, D. and Duriez, B. (1985) 'Local exchange and state intervention', in N. Redclift and E. Mingione (eds) *Beyond Employment: Household, Gender and Subsistence*, Oxford: Basil Blackwell.

Costes, L. (1991) 'Immigrés et travail clandestin dans le metro: les vendeurs a la sauvette', in S. Montagne-Villette (ed.) *Espaces et travail clandestins*, Paris: Masson.

Cousins, C. (1994) 'A comparison of the labour market position of women in Spain and the UK with reference to the "flexible" labour debate', *Work, Employment and Society* 8,1: 45–68.

Crompton, R., Gallie, D. and Purcell, K. (1996) 'Work, economic restructuring and social regulation', in R. Crompton, D. Gallie and K. Purcell (eds) *Changing Forms of Employment: Organisation, Skills and Gender*, London: Routledge.

Culpitt, I. (1992) *Welfare and Citizenship*, London: Sage.

Dagg, A. (1996) 'Organizing homeworkers into unions: the Homeworkers Association of Toronto, Canada', in E. Boris and E. Prugl (eds) *Homeworkers in Global Perspective: Invisible no More*, London: Routledge.

Dallago, B. (1991) *The Irregular Economy: The 'Underground' and the 'Black' Labour Market*, Aldershot: Dartmouth.

Daly, H.E. and Cobb, J.B. (1989) *For the Common Good: Redirecting the Economy*

Toward Community, the Environment and a Sustainable Future, London: Green Print.

Dasgupta, N. (1992) *Petty Trading in the Third World: The Case of Calcutta*, Aldershot: Avebury.

Dauncey, G. (1988) *After the Crash: The Emergence of the Rainbow Economy*, London: Green Print.

Davies, R.B., Elias, P. and Penn, R. (1992) 'The relationship between a husband's unemployment and his wife's participation in the labour force', *Oxford Bulletin of Economics and Statistics* 54,2: 145–71.

Dawes, L. (1993) *Long-term Unemployment and Labour Market Flexibility*, Leicester: Centre for Labour Market Studies, University of Leicester.

De Grazia, R. (1982) 'Clandestine employment: a problem for our time', in V. Tanzi (ed.) *The Underground Economy in the United States and Abroad*, Lexington, MA: Lexington Books.

—— (1984) *Clandestine Employment*, Geneva: International Labour Office.

De Klerk, L. and Vijgen, J. (1985) 'Cities in post-industrial perspective: new economies, new life-styles – new chances?', in ASVS (ed.) *Grote steden: verval of innovatie*, Lustrum: ASVS Lustrum Congress.

De Pardo, M.L., Castano, G.M. and Soto, A.T. (1989) 'The articulation of formal and informal sectors in the economy of Bogota, Colombia', in A. Portes, M. Castells and L. Benton (eds) *The Informal Economy: Studies in Advanced and Less Developed Countries*, Baltimore: Johns Hopkins University Press.

De Soto, H. (1989) *The Other Path*, London: I.B. Taurus.

Deakin, S. and Wilkinson, F. (1991/2) 'Social policy and economic efficiency: the deregulation of labour markets in Britain', *Critical Social Policy* 33: 40–51.

Dean, H. and Melrose, M. (1996) 'Unravelling citizenship: the significance of social security benefit fraud', *Critical Social Policy* 16: 3–31.

Del Boca, D. and Forte, F. (1982) 'Recent empirical surveys and theoretical interpretations of the parallel economy', in V. Tanzi (ed.) *The Underground Economy in the United States and Abroad*, Lexington, MA: Lexington Books.

Delphy, C. (1984) *Close to Home*, London: Hutchinson.

Denison, E. (1982) 'Is US growth understated because of the underground economy? Employment ratios suggest not', *Review of Income and Wealth* 28: 1–16.

Desproges, J. (1996) 'Making ecologically beneficial activities economically viable', in OECD (ed.) *Reconciling Economy and Society: Towards a Plural Economy*, Paris: OECD.

Dex, S. and Taylor, M. (1994) 'Household employment in 1991', *Employment Gazette* October: 353–7.

DfEE (1996) *Jobs: How the UK Compares*, London: DfEE.

Dicken, P. (1992) *Global Shift: The Internationalization of Economic Activity*, London: Paul Chapman.

Dilnot, A. (1992) 'Social security and labour market policy', in I.E. McLaughlin (ed.) *Understanding Employment*, London: Routledge.

Dilnot, A. and Morris, C.N. (1981) 'What do we know about the black economy?', *Fiscal Studies* 2: 58–73.

Dobson, R.V.G. (1993) *Bringing the Economy Home from the Market*, Montreal: Black Rose Books.

Dore, R. (1996) 'A feasible Jerusalem?', *The Political Quarterly* 67,1: 56–7.

Dorling, D. and Woodward, R. (1996) 'Social polarisation 1971–1991: a micro-geographical analysis of Britain', *Progress in Planning* 45,2: 67–122.

Douthwaite, R. (1996) *Short Circuit: Strengthening Local Economies for Security in an Unstable World*, Dartington: Green Books.

Duchrow, U. (1995) *Alternatives to Global Capitalism: Drawn from Biblical History, Designed for Political Action*, Oxford: Jon Carpenter Publishing.

Duncan, C.M. (1992) 'Persistent poverty in Appalachia: scarce work and rigid stratification', in C. Duncan (ed.) *Rural Poverty in America*, New York: Auburn House.

Economist Intelligence Unit (1982) *Coping with Unemployment: The Effects on the Unemployed Themselves*, London: Economist Intelligence Unit.

Ekins, P. (1992) *A New World Order: Grass Roots Movements for Global Change*, London: Routledge.

Elkin, T. and McLaren, D. (1991) *Reviving the City: Towards Sustainable Urban Development*, London: Friends of the Earth.

Engbersen, G., Schuyt, K., Timmer, J. and van Waarden, F. (1993) *Cultures of Unemployment: A Comparative Look at Long-term Unemployment and Urban Poverty*, San Francisco: Westview.

Espenshade, T.J. (1995a) 'Unauthorised immigration to the United States', *Annual Review of Sociology* 21: 195–216.

—— (1995b) 'Using INS Border Apprehension data to measure the flow of un-documented migrants across the US–Mexico frontier', *International Migration Review* 29,2: 545–65.

Esping-Andersen, G. (1994) 'Welfare states and the economy', in N.J. Smelser and R. Swedberg (eds) *The Handbook of Economic Sociology*, Princeton: Princeton University Press.

—— (1996) 'After the golden age? Welfare state dilemmas in a global economy', in G. Esping-Anderson (ed.) *Welfare States in Transition: National Adaptations in Global Economies*, London: Sage.

Etzioni, A. (1993) *The Spirit of Community: The Reinvention of American Society*, London: Simon and Schuster.

European Commission (1974) *Program of Action in Favour of Migrant Workers and their Families*, Luxembourg: Office for Official Publications of the European Communities.

— (1990) *Underground Economy and Irregular Forms of Employment, Final Synthesis Report*, Brussels: Office for Official Publications of the European Communities.

—— (1991) *Employment in Europe*, Luxembourg: Office for Official Publications of the European Communities.

—— (1993) *Employment in Europe*, Luxembourg: Office for Official Publications of the European Communities.

—— (1994) *The Demographic Situation in the European Union*, Luxembourg: Office for Official Publications of the European Communities.

—— (1995a) *The European Employment Strategy: Recent Progress and Prospects for the Future*, Luxembourg: Office for Official Publications of the European Communities.

—— (1995b) *Social Protection in Europe 1995*, Luxembourg: Office for Official Publications of the European Communities.

—— (1996a) *Employment in Europe 1996*, Luxembourg: European Commission DG for Employment, Industrial Relations and Social Affairs.

—— (1996b) *For a Europe of Civic and Social Rights: Report by the Comité des Sages*, Luxembourg: European Commission DG for Employment, Industrial Relations and Social Affairs.

Eurostat (1996) *1996 Facts through Figures: A Statistical Portrait of the European Union*, Brussels: Eurostat

Evason, E. and Woods, R. (1995) 'Poverty, deregulation of the labour market and benefit fraud', *Social Policy and Administration* 29,1: 40–55.

Evers, A. and Wintersberger, H. (1988) (eds) *Shifts in the Welfare Mix: Their Impact on Work, Social Services and Welfare Policies*, Vienna: European Centre for Social Welfare Training and Research.

Fainstein, N. (1996) 'A note on interpreting American poverty', in E. Mingione (ed.) *Urban Poverty and the Underclass*, Oxford: Basil Blackwell.

Farley, R. and Frey, W.H. (1994) 'Changes in the segregation of whites from blacks during the 1980s: small steps towards a more integrated society', *American Sociological Review* 59: 23–45.

Fashoyin, T. (1993) 'Nigeria: consequences for employment', in A. Adepoju (ed.) *The Impact of Structural Adjustment on the Population of Africa*, London: James Currey.

Feige, E.L. (1979) 'How big is the irregular economy?' *Challenge* November/ December: 5–13.

—— (1990) 'Defining and estimating underground and informal economies', *World Development* 18,7: 989–1002.

Feige, E.L. and McGee, R.T. (1983) 'Sweden's Laffer Curve: taxation and the unobserved economy', *Scandinavian Journal of Economics* 85,4: 499–519.

Felt, L.F. and Sinclair, P.R. (1992) '"Everyone does it": unpaid work in a rural peripheral region', *Work Employment and Society* 6,1: 43–64.

Ferman, L., Berndt, L. and Selo, E. (1978) *Analysis of the Irregular Economy: Cash-flow in the Informal Sector*, Detroit: Institute of Labour and Industrial Relations, University of Michigan/Wayne State University.

Fernandez-Kelly, M.P. and Garcia, A.M. (1989) 'Informalisation at the core: Hispanic women, homework, and the advanced capitalist state', *Environment and Planning D* 8: 459–83.

Fitchen, J.M. (1981) *Poverty in Rural America: A Case Study*, Boulder, CO: Westview.

Fortin, B. and Frechette, P. (1986) *Enquête sur les incidences et les perceptions de la taxation dans la region de Quebec*, Laval: Université de Laval.

—— (1987) *The Size and Determinants of the Underground Economy in Quebec*, Laval: Presses de l'Université de Laval.

Fortin, B., Garneau, G., Lacroix, G., Lemieux, T. and Montmarquette, C. (1996) *L'Economie souterraine au Quebec: mythes et réalités*, Laval: Presses de l'Université Laval.

Foudi, R., Stankiewicz, F. and Vanecloo, N. (1982) 'Chomeurs et economie informelle', *Cahiers de l'observation du changement social et culturel*, no. 17, Paris: CNRS.

Frank, A.G. (1996) 'The underdevelopment of development', in S.C. Chew and R.A. Denemark (eds) *The Underdevelopment of Development*, London: Sage.

Freud, D. (1979) 'A guide to underground economics', *Financial Times* 9 April: 16.

Frey, B.S. and Weck, H. (1983) 'What produces a hidden economy? An international cross-section analysis, *Southern Economic Journal* 49: 822–32.

Friedmann, Y. (1982) 'L'art de la survie' *Autogestions* 8–9: 89–95.

Gabor, I.R. (1988) 'Second economy and socialism: the Hungarian experience', in E.L. Feige (ed.) *The Underground Economies*, Cambridge: Cambridge University Press.

Gallie, D. (1985) 'Directions for the future', in B. Roberts, R. Finnegan and D. Gallie (eds) *New Approaches to Economic Life: Economic Restructuring, Unemployment and Social Divisions of Labour*, Oxford: Oxford University Press.

Galster, G.C. (1991) 'Housing discrimination and urban poverty of African-Americans', *Journal of Housing Research* 2: 87–122.

Gass, R. (1996) 'The next stage of structural change: towards a decentralised economy and an active society', in OECD (ed.) *Reconciling Economy and Society: Towards a Plural Economy*, Paris: OECD.

Geeroms, H. and Mont, J. (1987) 'Evaluation de l'importance de l'economie souterraine en Belgique: application de la methode monetaire', in V. Ginsburgh and P. Pestieau (eds) *L'Economie informelle*, Brussels: Editions Labor.

Gershuny, J. (1985) 'Economic development and change in the mode of provision of services', in N. Redclift and E. Mingione (eds) *Beyond Employment: Household, Gender and Subsistence*, Oxford: Basil Blackwell.

—— (1992) 'Change in the domestic division of labour in the UK, 1975–87: dependent labour versus adaptive partnership', in N. Abercrombie and A. Warde (eds) *Social Change in Contemporary Britain*, Cambridge: Polity.

Gershuny, J., Godwin, M. and Jones, S. (1994) 'The domestic labour revolution: a process of lagged adaptation', in M. Anderson, F. Bechhofer and J. Gershuny (eds) *The Social and Political Economy of the Household*, Oxford: Oxford University Press.

Gershuny, J. and Jones, S. (1987) 'The changing work/leisure balance in Britain 1961–84', *Sociological Review Monograph* 33,1: 9–50.

Gershuny, J., Miles, I., Jones, S., Mullins, C., Thomas, G. and Wyatt, S.M.E. (1986) 'Preliminary analyses of the 1983/4 ESRC Time Budget Data', *Quarterly Journal of Social Affairs* 2: 13–39.

Ghavamshahidi, Z. (1996) '"Bibi Khanum": carpet weavers and gender ideology in Iran', in E. Boris and E. Prugl (eds) *Homeworkers in Global Perspective: Invisible no More*, London: Routledge.

Giddens, A. (1984) *The Constitution of Society: Outline of the Theory of Structuration*, Cambridge: Polity.

Gilbert, A. (1994) 'Third World Cities: poverty, unemployment, gender roles and the environment during a time of restructuring', *Urban Studies* 31, 4/5: 605–33.

Gilder, G. (1981) *Wealth and Poverty*, New York: Basic Books.

Ginsburgh, V., Perelman, S. and Pestieau, P. (1987) 'Le travail au noir', in V. Ginsburgh and P. Pestieau (eds) *L'Economie informelle*, Brussels: Editions Labor.

Glatzer, W. and Berger, R. (1988) 'Household composition, social networks and household production in Germany', in R.E. Pahl (ed.) *On Work: Historical, Comparative and Theoretical Approaches*, Oxford: Basil Blackwell.

Glennie, P.D. and Thrift, N.J. (1992) 'Modernity, urbanism and modern consumption', *Environment and Planning A* 10: 423–33.

Gornick, J.C. and Jacobs, J.A. (1996) 'A cross-national analysis of the wages of part-time workers: evidence from the United States, the United Kingdom, Canada and Australia', *Work, Employment and Society* 10,1: 1–27.

Gorz, A. (1985) *Paths to Paradise*, London: Pluto.

Gray, J. (1984) *Hayek on Liberty*, Oxford: Basil Blackwell.

Greco, T.H. (1994) *New Money for Healthy Communities*, Tucson, AR: T.H. Greco.

Greffe, X. (1981) 'L'economie non-officielle', *Consommation* 3: 5–16.

Gregg, P. and Wadsworth, J. (1996) *It Takes Two: Employment Polarisation in the OECD*, Discussion Paper no. 304, London: Centre for Economic Performance, London School of Economics.

Gregory, A. and Windebank, J. (forthcoming) *Women and Work in France and Britain*, Basingstoke: Macmillan.

Gregson, N. and Lowe, M. (1994) *Servicing the Middle Classes*, London: Routledge.

Gringeri, C. (1996) 'Making cadillacs and buicks for General Motors', in E. Boris and E. Prugl (eds) *Homeworkers in Global Perspective: Invisible no More*, London: Routledge.

Grosskoff, R. and Melgar, A. (1990) 'Sector informeal urbano: ingreso, empleo y demanda de su produccion: el caso Uruguayo', in PREALC (ed.) *Ventas informales: relaciones con el sector moderno*, Santiago: PREALC-OIT.

Grossman, G. (1989) 'Informal personal incomes and outlays of the Soviet urban population', in A. Portes, M. Castells, and L.A. Benton (eds) *The Informal Economy: Studies in Advanced and Less Developing Countries*, Baltimore: Johns Hopkins University Press.

Guisinger, S. and Irfan, M. (1980) 'Pakistan's informal sector', *Journal of Development Studies* 16,4: 412–26.

Gutmann, P.M. (1977) 'The subterranean economy', *Financial Analysts Journal* 34,11: 26–7.

—— (1978) 'Are the unemployed, unemployed?', *Financial Analysts Journal* 34,1: 26–9.

Hadjicostandi, J. (1990) 'Facon: women's formal and informal work in the garment industry in Kavala, Greece', in K. Ward (ed.) *Women Workers and Global Restructuring*, Ithaca, NY: ILR Press.

Hadjimichalis, C. and Vaiou, D. (1989) 'Whose flexibility?: The politics of informalisation in Southern Europe', Paper presented to the IAAD/SCG Study Groups of the IBG conference on *Industrial Restructuring and Social Change: The Dawning of a New Era of Flexible Accumulation?*, Durham.

Hahn, J. (1996) '"Feminization through flexible labor": the political economy of home-based work in India', in E. Boris and E. Prugl (eds) *Homeworkers in Global Perspective: Invisible no More*, London: Routledge.

Hakim, C. (1995) 'Five feminist myths about women's employment', *British Journal of Sociology* 46: 429–55.

Hall, P. (1996) 'The global city', *International Social Science Journal* 147: 15–24.

Hanson, S. and Pratt, G. (1995) *Gender, Work and Space*, London: Routledge.

Harding, P. and Jenkins, R. (1989) *The Myth of the Hidden Economy: Towards a New Understanding of Informal Economic Activity*, Milton Keynes: Open University Press.

Harris, J.R. and Todaro, M. (1970) 'Migration, unemployment and development: a two-sector analysis', *American Economic Review* 60: 126–42.

Hart, K. (1973) 'Informal income opportunities and urban employment in Ghana', *Journal of Modern African Studies* 11: 61–89.

—— (1990) 'The idea of economy: six modern dissenters', in R. Friedland and A.F. Robertson (eds) *Beyond the Marketplace: Rethinking Economy and Society*, New York: Aldine de Gruyter.

Hasseldine, D. and Bebbington, K. (1991) 'Blending economic deterrence and fiscal psychology models in the design of response to tax evasion: the New Zealand experience', *Journal of Economic Psychology* 12,2: 2–19.

Haughton, G., Johnson, S., Murphy, L. and Thomas, K. (1993) *Local Geographies of Unemployment: Long-term Unemployment in Areas of Local Deprivation*, Aldershot: Avebury.

Heinze, R.G. and Olk, T. (1982) 'Development of the informal economy: a strategy for resolving the crisis of the welfare state', *Futures* June: 189–204.

Hellberger, C. and Schwarze, J. (1986) *Umfang und struktur der nebenerwerbstatigkeit in der Bundesrepublik Deutschland*, Berlin: Mitteilungen aus der Arbeitsmarket- und Berufsforschung.

—— (1987) 'Nebenerwerbstatigkeit: ein indikator fur arbeitsmarkt-flexibilitat oder schattenwirtschaft', *Wirtschaftsdienst* 2: 83–90.

Henry, J. (1976) 'Calling in the big bills', *Washington Monthly* 5: 6.

Henry, S. (1978) *The Hidden Economy*, London: Martin Robertson.

Herbert, A. and Kempson, E. (1996) *Credit Use and Ethnic Minorities*, London: Policy Studies Institute.

Hessing, D., Elffers, H., Robben, H. and Webley, P. (1993) 'Needy or greedy? The social pyschology of individuals who fraudulently claim unemployment benefits', *Journal of Applied Social Psychology* 23,3: 226–43.

Hewitt, P. (1996) 'The place of part-time employment' in P. Meadows (ed.) *Work Out or Work In? Contributions to the Debate on the Future of Work*, York: Joseph Rowntree Foundation.

HM Treasury (1994) 'Developments in the personal sector', *Economic Briefing* 7: 9–11.

Holzer, H. (1991) 'The spatial mismatch hypothesis: what has the evidence shown?', *Urban Studies* 28: 105–22.

Houghton, D. (1979) 'The futility of taxation menaces', in A. Seldon (ed.) *Tax Avoision*, London: Institute of Economic Affairs.

Houston, J.F. (1990) 'The policy implications of the underground economy', *Journal of Economics and Business* 42,1: 27–37.

Howe, L. (1988) 'Unemployment, doing the double and local labour markets in

Belfast', in C. Cartin and T. Wilson (eds) *Ireland from Below: Social Change and Local Communities in Modern Ireland*, Dublin: Gill and Macmillan.

—— (1990) *Being Unemployed in Northern Ireland: An Ethnographic Study*, Cambridge: Cambridge University Press.

Hoyman, M. (1987) 'Female participation in the informal economy: a neglected issue', *Annals of the American Academy of Political and Social Science* 493: 64–82.

Huber, E. (1996) 'Options for social policy in Latin America: neo-liberal versus social democratic models', in G. Esping-Anderson (ed.) *Welfare States in Transition: National Adaptations in Global Economies*, London: Sage.

Hutton, W. (1995) *The State We're In*, London: Vintage.

International Labour Organisation (1972) *Employment, Incomes and Equality: A Strategy for Increasing Productive Employment in Kenya*, Geneva: International Labour Office.

—— (1996) *World Employment 1996/97: National Policies in a Global Context*, Geneva: International Labour Office.

Isachsen, A.J., Klovland, J.T. and Strom, S. (1982) 'The hidden economy in Norway', in V. Tanzi (ed.) *The Underground Economy in the United Sates and Abroad*, Lexington, MA: D.C. Heath.

Isachsen, A.J. and Strom, S. (1985) 'The size and growth of the hidden economy in Norway', *Review of Income and Wealth* 31,1: 21–38.

Isuani, E.A. (1985) 'Social security and public assistance', in C. Mesa-Lago (ed.) *The Crisis of Social Security and Health Care: Latin American Experiences and Lessons*, Latin American Monograph and Document Series no. 9, Pittsburgh: Center for Latin American Studies, University of Pittsburgh.

Izquierdo, M.J., Miguelez, F. and Subirats, M. (1987) *Enquesta Metropolitana, Volume I: Informe general*, Barcelona: Instituto de Estudios Metropolitanos de Barcelona.

Jackson, K.T. (1995) *Crabgrass Frontier: The Suburbanisation of the United States*, Oxford: Oxford University Press.

Jarvis, H. (1997) 'Housing, labour markets and household structure: questioning the role of secondary data analysis in sustaining the polarization debate', *Regional Studies* 31,5: 521–31.

Jencks, C. and Mayer, S. (1990) 'Residential segregation, job proximity and black job opportunities: the empirical status of the spatial mismatch hypothesis', in M. McGreary (ed.) *Poverty in America*, Washington, DC: National Academy Press.

Jensen, L., Cornwell, G.T. and Findeis, J.L. (1995) 'Informal work in nonmetropolitan Pennsylvania', *Rural Sociology* 60,1: 91–107.

Jessen, J., Siebel, W., Siebel-Rebell, C., Walther, U. and Weyrather, I. (1987) 'The informal work of industrial workers', Paper presented at *6th Urban Change and Conflict Conference*, University of Kent at Canterbury, September.

Johnson-Anumonwo, I., McLafferty, S. and Preston, V. (1994) 'Gender, race and the spatial concentration of women's employment', in J. Garber and R. Turner (eds) *Gender in Urban Research*, Thousand Oaks, CA: Sage.

Jones, P.N. (1994) 'Economic restructuring and the role of foreign workers in the 1980s: the case of Germany', *Environment and Planning A* 26: 1435–53.

Jones, T. (1993) *Britain's Ethnic Minorities*, London: Policy Studies Institute.

Jordan, B., James, S., Kay, H. and Redley, M. (1992) *Trapped in Poverty*, London: Routledge.

Jordan, B. and Redley, M. (1994) 'Polarisation, underclass and the welfare state', *Work, Employment and Society* 8,2: 153–76.

Jung, Y.H., Snow, A. and Trandel, G.A. (1994) 'Tax evasion and the size of the underground economy', *Journal of Public Economics* 54: 391–402.

Kain, J. (1968) 'Housing segregation, negro employment and metropolitan decentralization', *Quarterly Journal of Economics* 82,2: 175–97.

Kalinda, B. and Floro, M. (1992) *Zambia in the 1980s: A Review of National and Urban Level Economic Reforms*, Working Paper 18, Washington, DC: Urban Development Division, World Bank.

Kasarda, J.D. (1989) 'Urban industrial transition and the urban underclass', *Annals of the American Academy of Political and Social Science* 501: 26–47.

Keenan, A. and Dean, P.N. (1980) 'Moral evaluation of tax evasion', *Social Policy and Administration* 14: 209–20.

Kesteloot, C. and Meert, H. (1994) 'Les fonctions socio-economiques de l'economie informelle et son implantation spatiale dans les villes belges', Paper presented to the international conference on *Cities, Enterprises and Society at the Eve of the XXIst Century*, Lille, March.

Kiernan, K. (1992) 'The roles of men and women in tomorrow's Europe', *Employment Gazette* October: 491–8.

Kinsey, K. (1992) 'Deterrence and alienation effects of IRS enforcement: an analysis of survey data', in J. Slemrod (ed.) *Why People Pay Taxes*, Michigan: University of Michigan Press.

Klovland, J.T. (1980) *In Search of the Hidden Economy: Tax Evasion and the Demand for Currency in Norway and Sweden*, Bergen: Norwegian School of Economics and Business Administration.

Komter, A.E. (1996) 'Reciprocity as a principle of exclusion: gift giving in the Netherlands', *Sociology* 30,2: 299–316.

Koopmans, C.C. (1989) *Informele Arbeid: vraag, aanbod, participanten, prijzen*, Amsterdam: Proefschrift Universitiet van Amsterdam.

Korbonski, A. (1981) 'The "second economy" in Poland', *Journal of International Affairs* 35,1: 1–12.

Kroft, H.G., Engbersen, G., Schuyt, K. and Van Waarden, F. (1989) *Een Tijd Zonder Werk: een onderzoek naar de levenswereld van langdurig werklozen*, Leiden: University of Leiden.

Kumar, K. (1978) *Prophecy and Progress: The Sociology of Industrial and Post-industrial Society*, London: Allen Lane.

Lagos, R.A. (1995) 'Formalising the informal sector: barriers and costs', *Development and Change* 26: 110–31.

Laguerre, M.S. (1994) *The Informal City*, London: Macmillan.

Lang, P. (1994) *LETS Work*, Bristol: Grover Books.

Lautier, B. (1990) 'Wage relationship, formal sector and employment policy in South America', *Journal of Development Studies* 26: 278–98.

—— (1994) *L'Economie informelle dans le tiers monde*, Paris: La Decouverte.

Laville, J-L. (1995) 'La crise de la condition salariale: emploi, activité et nouvelle question sociale', *Esprit*, 12,12: 32–54.

—— (1996) 'Economy and solidarity: exploring the issues', in OECD (ed.) *Reconciling Economy and Society: Towards a Plural Economy*, Paris: OECD.

Lee, R. (1996) 'Moral money? LETS and the social construction of local economic geographies in Southeast England', *Environment and Planning A* 28: 1377–94.

Legrain, C. (1982) 'L'economie informelle a Grand Failly', *Cahiers de l'OCS*, no. 7, Paris: CNRS.

Lemieux, T., Fortin, B. and Frechette, P. (1994) 'The effect of taxes on labor supply in the underground economy', *American Economic Review* 84,1: 231–54.

Leonard, M. (1994) *Informal Economic Activity in Belfast*, Aldershot: Avebury.

Leontidou, L. (1993) 'Informal strategies of unemployment relief in Greek cities: the relevance of family, locality and housing', *European Planning Studies*, 1,1: 43–68.

Levitan, L. and Feldman, S. (1991) 'For love or money: non-monetary economic arrangements among rural households in central New York', *Research in Rural Sociology and Development* 5: 149–72.

Lewis, A. (1959) *The Theory of Economic Growth*, London: Allen and Unwin.

Leyshon, A. and Thrift, N. J. (1994) 'Geographies of financial exclusion: financial abandonment in Britain and the United States', *Transactions of the Institute of British Geographers* 20: 312–41.

Lim, I.Y.C. (1993) 'Capitalism, imperialism and patriarchy: the dilemma of third-world women workers in multinational factories', in J. Nash and M.P. Fernandez-Kelly (eds) *Women, Men and the International Division of Labour*, Albany, NY: State University of New York Press.

Lin, J. (1995) 'Polarized development and urban change in New York's China-town', *Urban Affairs Review*, 30,3: 332–54.

Lindbeck, A. (1981) *Work Disincentives in the Welfare State*, Stockholm: Institute for International Economic Studies, University of Stockholm.

Linton, M. (1986) 'Local currency', in P. Ekins (ed.) *The Living Economy: A New Economics in the Making*, London: Routledge.

Lipietz, A. (1992) *Towards a New Economic Order: Post-Fordism, Ecology and Democracy*, Cambridge: Polity.

—— (1995) *Green Hopes: The Future of Political Ecology*, Cambridge: Polity.

Lobo, F.M. (1990a) 'Irregular work in Spain', in European Commission, *Underground Economy and Irregular Forms of Employment, Final Synthesis Report*, Brussels: Office for Official Publications of the European Communities.

—— (1990b) 'Irregular work in Portugal', in European Commission, *Underground Economy and Irregular Forms of Employment, Final Synthesis Report*, Brussels: Office for Official Publications of the European Communities.

Lomnitz, L.A. (1988) 'Informal exchange networks in formal systems: a theoretical model', *American Anthropologist* 90: 42–55.

Lopez, C. (1986) *El Textil Irregular en Terrassa 1975–1985*, Terrassa: Ajuntament de Terrassa.

Lozano, B. (1985) *High Technology, Cottage Industry*, PhD dissertation, Davis, CA: Department of Sociology, University of California.

—— (1989) *The Invisible Workforce: Transforming American Business with Outside and Home-based Workers*, New York: Free Press.

Lubell, H. (1991) *The Informal Sector in the 1980s and 1990s*, Paris: OECD.

Lysestol, P.M. (1995) '"The other economy" and its influences on job-seeking behaviour for the long-term unemployed', Paper presented at the *Euroconference on Social Policy in an Environment of Insecurity: Contemporary Dilemmas and Challenges for Social Policy*, Lisbon, November.

Mabogunje, A.L. and Filani, M.O. (1981) 'The informal sector in a small city: the case of Kano, Nigeria', in S.V. Setheraman (ed.) *The Urban Informal Sector in Developing Countries: Employment, Poverty and Environment*, Geneva: International Labour Office.

Macafee, K. (1980) 'A glimpse of the hidden economy in the national accounts' *Economic Trends* 2: 81–7.

McCrohan, K., Smith, J.D. and Adams, T.K. (1991) 'Consumer purchases in informal markets: estimates for the 1980s, prospects for the 1990s', *Journal of Retailing* 67: 22–50.

MacDonald, R. (1994) 'Fiddly jobs, undeclared working and the something for nothing society', *Work, Employment and Society* 8,4: 507–30.

Macfarlane, R. (1997) *Unshackling the Poor: A Complementary Approach Towards Local Economic Development*, York: Joseph Rowntree Foundation.

McInnis-Dittrich, K. (1995) 'Women of the shadows: Appalachian women's participation in the informal economy', *Affilia: Journal of Women and Social Work* 10,4: 398–412.

McLafferty, S. and Preston, V. (1996) 'Spatial mismatch and employment in a decade of restructuring', *The Professional Geographer* 48,4: 420–31.

McLaughlin, E. (1994) *Flexibility in Work and Benefits*, London: Institute of Public Policy Research.

Maguire, K. (1993) 'Fraud, extortion and racketeering: the black economy in Northern Ireland', *Crime, Law and Social Change* 20: 273–92.

Main, B.G.M. (1994) 'The labour market: friend or foe?', in M. Anderson, F. Bechhofer and J. Gershuny (eds) *The Social and Political Economy of the Household*, Oxford: Oxford University Press.

Maldonado, C. (1995) 'The informal sector: legalization or laissez-faire?' *International Labour Review* 134,6: 705–28.

Malone, A. (1994) 'A beggar's banquet', *Sunday Times* 20 February: 1.14.

Marsden, T. and Wrigley, N. (1994) 'Regulation, retailing and consumption', Paper presented to the *Annual Meeting of the Association of American Geographers*, San Francisco, March.

Martin, C.J. (1996) 'Economic strategies and moral principles in the survival of poor households in Mexico', *Bulletin of Latin American Research* 15,2: 193–210.

Martin, J. and Roberts, C. (1984) *Women and Employment: A Lifetime Perspective*, London: Department of Employment and Office of Population, Censuses and Surveys.

Martino, A. (1981) 'Measuring Italy's underground economy', *Policy Review* Spring: 87–106.

Massey, D. (1984) *Spatial Divisions of Labour: Social Structures and the Geography of Production*, Basingstoke: Macmillan.

Massey, D.S. and Denton, N.A. (1993) *American Apartheid*, Cambridge, MA: Harvard, University Press.

Mathur, O.P. and Moser, C. (1984) 'The urban informal sector: an agenda for research', *Regional Development Dialogue* 5: ix–xxi.

Mattera, P. (1980) 'Small is not beautiful: decentralised production and the underground economy in Italy', *Radical America* 14,5: 67–76.

—— (1985) *Off the Books: The Rise of the Underground Economy*, New York: St Martin's Press.

Matthews, K.G.P. (1982) 'Demand for currency and the black economy in the UK', *Journal of Economic Studies* 9,2: 3–22.

—— (1983) 'National income and the black economy', *Journal of Economic Affairs* 3,4: 261–67.

Matthews, K.G.P. and Rastogi, A. (1985) 'Little mo and the moonlighters: another look at the black economy', *Quarterly Economic Bulletin* 6: 21–4.

Mayo, E. (1996) 'Dreaming of work', in P. Meadows (ed.) *Work Out or Work In? Contributions to the Debate on the Future of Work*, York: Joseph Rowntree Foundation.

Meade, J.E. (1995) *Full Employment Regained? An Agathopian Dream*, Cambridge: Cambridge University Press.

—— (1996) 'Full employment, new technologies and the distribution of income', in M. Bulmer and A.M. Rees (eds) *Citizenship Today: The Contemporary Relevance of T.H. Marshall*, London: UCL Press.

Meadows, T.C. and Pihera, J.A. (1981) 'A regional perspective on the underground economy', *Review of Regional Studies* 11: 83–91.

Meagher, K. (1995) 'Crises, informalization and the urban informal sector in sub-Saharan Africa', *Development and Change* 26,2: 259–84.

Meehan, E. (1993) *Citizenship and the European Community*, London: Sage.

Meert, H., Mestiaen, P. and Kesteloot, C. (1997) 'The geography of survival: household strategies in urban settings', *Tijdschrift voor Economische en Sociale Geografie* 88,2: 169–81.

Miguelez, F. and Recio, A. (1986) 'Catalunya: la economia ignota', in Institut Alfons el Magnanim (ed.) *Economia Sumergida en Espana*, Valencia: Institut Alfons el Magnanim.

Miles, I. (1983) *Adaptation to Unemployment?*, Occasional Paper no. 20, Brighton: Science Policy Research Unit, University of Sussex.

Minc, A. (1980) 'Le chomage et l'economie souterraine', *Le Debat* 2: 3–14.

—— (1982) *L'Après-Crise a commencé*, Paris: Gallimard.

Mingione, E. (1990) 'The case of Greece', in European Commission, *Underground Economy and Irregular Forms of Employment, Final Synthesis Report*, Brussels: Office for Official Publications of the European Communities.

—— (1991) *Fragmented Societies: A Sociology of Economic Life Beyond the Market Paradigm*, Oxford: Basil Blackwell.

—— (1994) 'Socio-economic restructuring and social exclusion', Paper presented to the international conference on *Cities, Enterprises and Society at the Eve of the XXIst century*, Lille, March.

Mingione, E. and Magatti, M. (1995) *Social Europe Follow up to the White Paper: The Informal Sector*, DG V Supplement 3/95, Brussels: European Commission.

Mingione, E. and Morlicchio, E. (1993) 'New forms of urban poverty in Italy: risk path models in the North and South', *International Journal of Urban and Regional Research* 17,3: 413–27.

Miraftab, F. (1996) 'Space, gender and work: home-based workers in Mexico', in E. Boris and E. Prugl (eds) *Homeworkers in Global Perspective: Invisible no More*, London: Routledge.

Mirus, R. and Smith, R.S. (1989) 'Canada's underground economy', in E.L. Feige (ed.) *The Underground Economies: Tax Evasion and Information Distortion*, Cambridge: Cambridge University Press.

Modood, T. and Berthoud, R. (1997) *Ethnic Minorities in Britain: Diversity and Disadvantages*, London: Policy Studies Institute.

Mogensen, G.V. (1985) *Sort Arbejde i Danmark*, Copenhagen: Institut for Nationaløkonomi.

—— (1990) 'Black markets and welfare in Scandinavia: some methodological and empirical issues', in M. Estellie Smith (ed.) *Perspectives on the Informal Economy*, New York: University Press of America.

Mogensen, G.V., Kvist, H.K., Kormendi, E. and Pedersen, S. (1995) *The Shadow Economy in Denmark 1994: Measurement and Results*, Copenhagen: Rockwool Foundation Research Unit.

Momsen, J.H. (1991) *Women and Development in the Third World*, London: Routledge.

Morris, L. (1987) 'Constraints on gender: the family wage, social security and the labour market; reflections on research in Hartlepool', *Work, Employment and Society* 1,1: 85–106.

—— (1988) 'Employment, the household and social networks', in D. Gallie (ed.) *Employment in Britain*, Oxford: Basil Blackwell.

—— (1993) 'Is there a British underclass?', *International Journal of Urban and Regional Research* 17,3: 404–12.

—— (1994) 'Informal aspects of social divisions', *International Journal of Urban and Regional Research* 18: 112–26.

—— (1995) *Social Divisions: Economic Decline and Social Structural Change*, London: UCL Press.

Moulier-Boutang, Y. (1991) 'Dynamique des migrations internationales et economie souterraine: compraison internationale et perspectives européenes', in S. Montagne-Villette (ed.) *Espaces et travail clandestins*, Paris: Masson,

Murray, C. (1984) *Losing Ground: American Social Policy, 1950–1980*, New York: Basic Books.

Myles, J. (1996) 'When markets fail: social welfare in Canada and the US', in G. Esping–Anderson (ed.) *Welfare States in Transition: National Adaptations in Global Economies*, London: Sage.

Nattrass, N.J. (1987). 'Street trading in Transkei – a struggle against poverty, persecution and prosecution', *World Development* 15,7: 861–75.

Nicaise, I. (1996) *Which Partnerships for Employment? Social Partners, NGOs and Public Authorities*, European Social Policy Forum Working Paper II, Brussels: European Commission, DG V.

Noble, M. and Turner, R. (1985) *The Moral Implications of Unemployment and the Hidden Economy in a Scottish Village*, Swindon: ESRC End of Award Report, award no. G00232114.

Nurulk Amin, A.T.M. (1987) 'The role of the informal sector in economic development: some evidence from Dhaka, Bangladesh', *International Labour Review* 126,5: 611–23.

Nyssens, M. (1996) 'Popular economy in the south, third sector in the North, seeds of a mutually supportive sector?', in OECD (ed.) *Reconciling Economy and Society: Towards a Plural Economy*, Paris: OECD.

Oakley, A. and Williams, S. (1994) (eds) *The Politics of the Welfare State*, London: UCL Press.

OECD (Organisation for Economic Co-operation and Development) (1993) *Employment Outlook*, Paris: OECD.

—— (1994) *Jobs Study: Part 2*, Paris: OECD.

—— (1996) (ed.) *Reconciling Economy and Society: Towards a Plural Economy*, Paris: OECD.

Offe, C. and Heinze, R.G. (1992) *Beyond Employment: Time, Work and the Informal Economy*, Cambridge: Polity.

O'Higgins, M. (1981) 'Tax evasion and the self-employed', *British Tax Review* 26: 367–78.

Okun, A.M. (1975) *Equality and Efficiency: The Big Trade-off*, Washington, DC: Brookings Institute.

Oliver, M. and Shapiro, T.M. (1997) *Black Wealth/White Wealth: A New Perspective on Racial Inequality*, London: Routledge.

OPCS (Office of Population, Censuses and Surveys) (1992) *General Household Survey: Carers in 1990*, OPCS Monitor ss92/2, London: OPCS.

Owen, D. (1994) *Ethnic Minority Women and the Labour Market: An Analysis of the 1991 Census*, Manchester: Equal Opportunities Commission.

Pacione, M. (1997) 'Local Exchange Trading Systems as a response to the globalisation of capitalism', *Urban Studies* 34,8: 1179–99.

Paglin, M. (1994) 'The underground economy: new estimates from household income and expenditure surveys', *The Yale Law Journal* 103,8: 2239–57.

Pahl, R.E. (1984) *Divisions of Labour*, Oxford: Basil Blackwell.

—— (1985a) 'The restructuring of capital, the local political economy and household work strategies', in D. Gregory and J. Urry (eds) *Social Relations and Spatial Structures*, London: Macmillan.

—— (1985b) 'The politics of work', *The Political Quarterly* 56,4: 331–45.

—— (1987) 'Does jobless mean workless? Unemployment and informal work', *Annals of the American Academy of Political and Social Science* 493: 36–46.

—— (1988) 'Some remarks on informal work, social polarization and the social structure', *International Journal of Urban and Regional Research* 12, 2: 247–67.

—— (1990) 'The black economy in the United Kingdom', in *European Commission, Underground Economy and Irregular Forms of Employment, Final Synthesis Report*, Brussels: Office for Official Publications of the European Communities.

Paredes-Cruzatt, P. (1987a) *Segmentacion del mercado laborel em Lima Metropolitana*, Lima: Planificacion del Mercado Laboral, Projecto Per/85/007, Cuader-no de Informaciones no. 1

—— (1987b) *Condiciones de trabajo de los trabajadores del sector informal urbano de Lima Metropolitana*, Lima: Planificacion del Mercado Laboral, Projecto Per/ 85/007, Cuader-no de Informaciones no. 4.

Parker, H. (1982) 'Social security foments the black economy', *Economic Affairs* 3: 32–5.

—— (1989) *Instead of the Dole: An Enquiry into Integration of the Tax and Benefit Systems*, London: Routledge.

Peacock, A.T. and Shaw, G.K. (1982) 'Tax evasion and tax revenue loss', *Public Finance* 37: 269–78.

Peattie, L.R. (1980) 'Anthropological perspectives on the concepts of dualism, the informal sector and marginality in developing urban economies', *International Regional Science Review* 5,1: 1–31.

Peck, J. (1996a) *Work-Place: The Social Regulation of Labour Markets*, London: Guildford Press.

—— (1996b) *The Geo-Politics of the Workfare State*, Paper presented at the *Centre for Urban Development and Environmental Management Seminar*, Leeds Metropolitan University, October.

Pestieau, P. (1984) *Belgium's Irregular Economy*, Brussels: ULB.

—— (1985) 'Belgium's irregular economy', in W. Gaeartner and A. Wenig (eds) *The Economics of the Shadow Economy*, Berlin: Springer-Verlag.

—— (1989) *L'Economie souterraine*, Brussels: Hachette.

Peterson, H.G. (1982) 'Size of the public sector, economic growth and the informal economy: development trends in the Federal Republic of Germany', *Review of Income and Wealth* 28: 191–215.

Phizacklea, A. (1990) *Unpacking the Fashion Industry: Gender, Racism and Class in Production*, London: Routledge.

Phizacklea, A. and Wolkowitz, C. (1995) *Homeworking Women: Gender, Racism and Class at Work*, London: Sage.

Pinch, S. (1994) 'Social polarization: a comparison of evidence from Britain and the United States', *Environment and Planning A* 25,6: 779–95.

—— (1997) *Worlds of Welfare: Understanding the Changing Geographies of Social Welfare Provision*, London: Routledge.

Pinch, S. and Storey, A. (1992) 'Who does what where? A household survey of the division of domestic labour in Southampton', *Area* 24: 5–12.

Polanyi, K. (1957) *The Great Transformation*, Boston: Beacon Press.

Porritt, J. (1996) 'Local jobs depend on local initiative', *Finance North* September/ October: 88.

Porter, R.D. and Bayer, A.S. (1989) 'Monetary perspective on underground economic activity in the United States', in E.L. Feige (ed.) *The Underground Economies: Tax Evasion and Information Distortion*, Cambridge: Cambridge University Press.

Portes, A. (1994) 'The informal economy and its paradoxes', in N.J. Smelser and R. Swedberg (eds) *The Handbook of Economic Sociology*, Princeton: Princeton University Press.

Portes, A., Blitzer, S. and Curtis, J. (1986) 'The urban informal sector in Uruguay: its internal structure, characteristics and effects', *World Development* 14,6: 727–41.

Portes, A. and Sassen-Koob, S. (1987) 'Making it underground: comparative material on the informal sector in Western market economies', *American Journal of Sociology* 93,1: 30–61.

Portes, A. and Stepick, A. (1993) *City on the Edge: The transformation of Miami*, Los Angeles: University of California Press.

Priest, G.L. (1994) 'The ambiguous moral foundations of the underground economy', *The Yale Law Journal* 103,8: 2259–88.

Prugl, E. (1996) 'Home-based producers in development discourse', in E. Boris and E. Prugl (eds) *Homeworkers in Global Perspective: Invisible no More*, London: Routledge.

Psacharopoulos, G. and Tzannatos, Z. (1992) *Women's Employment and Pay in Latin America: Overview and Methodology*, Washington, DC: World Bank Regional and Sectoral Studies.

Pugliese, E. (1994) 'The question of irregularly employed migrant work from less developed countries: recent trends and changes', in *Social Europe Follow-up to the White Paper: The Informal Sector, Supplement 3/95*, Brussels: European Commission DG V.

Rakowski, C.A. (1994) 'Convergence and divergence in the informal sector debate: a focus on Latin America, 1984–92', *World Development* 22,4: 501–16.

Recio, A. (1988) *El Trabajo precario en Catalunya: la industria textilanera des valles Occidental*, Barcelona: Commission Obrera Nacional de Catalunya.

Rehfeldt, U. (1992) 'Maastricht: l'extension des competences européennes en matière sociale', *Problèmes Economiques* 2248: 1–5.

Reissert, B. (1994) 'Unemployment compensation and the labour market: a European perspective', in S. Mangen and L. Hantrais (eds) *Unemployment, the Informal Economy and Entitlement to Benefit*, Loughborough: European Research Centre, University of Loughborough.

Renooy, P. (1984) *De schemerzone: 'werplaats' tussen vrije tijd en arbeid*, The Hague: Ministry of Social Affairs and Employment.

—— (1990) *The Informal Economy: Meaning, Measurement and Social Significance*, no. 115, Amsterdam: Netherlands Geographical Studies.

Richardson, H.W. (1984) 'The role of the informal sector in developing countries: an overview' *Regional Development Dialogue* 5,2: 3–55.

Rifkin, J. (1995) *The End of Work*, London: G.P. Putman's.

Roberts, B. (1989) 'Employment structure, life cycle and life chances: formal and informal sectors in Guadalajara', in A. Portes, M. Castells and L. Benton (eds) *The Informal Economy: Comparative Studies in Advanced and Third World Countries*, Baltimore: Johns Hopkins University Press.

—— (1990) 'The informal sector in comparative perspective', in M. Estellie Smith (ed.) *Perspectives on the Informal Economy*, New York: University Press of America.

—— (1991) 'Household coping strategies and urban poverty in a comparative perspective', in M. Gottdiener and C.G. Pickvance (eds) *Urban Life In Transition*, London: Sage.

—— (1992) *The Dynamics of Informal Employment in Mexico*, Discussion Paper Series on the Informal Sector no.3, Washington, DC: Bureau of International Labor Affairs, US Department of Labor.

—— (1994) *Minority Ethnic Women: Unemployment and Education*, Manchester: Equal Opportunities Commission.

Robertson, J. (1981) 'The future of work: some thoughts about the roles of men and women in the transition to a SHE future', *Women's Studies International Quarterly* 4,1: 83–94.

—— (1985) *Future Work: Jobs, Self-employment and Leisure after the Industrial Age*, Aldershot: Gower/Temple Smith.

—— (1991) *Future Wealth: A New Economics for the 21st Century*, London: Cassells.

—— (1994) *Taxes and Benefits*, London: New Economics Foundation.

—— (1996) 'Towards a new social compact: citizen's income and radical tax reform', *The Political Quarterly* 67,1: 54–5.

Robson, B.T. (1988) *Those Inner Cities: Reconciling the Social and Economic Aims of Urban Policy*, Oxford: Clarendon.

Roldan, M. (1985) 'Industrial outworking, struggles for the reproduction of working-class families and gender subordination', in N. Redclift and E. Mingione (eds) *Beyond Employment: Household, Gender and Subsistence*, Oxford: Basil Blackwell.

Rosanvallon, P. (1980) 'Le developpement de l'economie souterraine et l'avenir des sociétés industrielles', *Le Debat* 2: 8–23.

Rosewell, B. (1996) 'Employment, households and earnings', in P. Meadows (ed.) *Work Out or Work In? Contributions to the Debate on the Future of Work*, York: Joseph Rowntree Foundation.

Rostow, W.J. (1960) *The Stages of Economic Growth: A Non-communist Manifesto*, Cambridge: Cambridge University Press.

Roth, J., Scholz, J. and Witte, A. (1992) *Tax-payer Compliance, Volume 1: An Agenda for Research*, Pennsylvania: University of Pennsylvania Press.

Roustang, G. (1987) *L'Emploi: un choix de société*, Paris: Syros.

Rowlingson, K., Whyley, C., Newburn, T. and Berthoud, R. (1997) *Social Security Fraud*, DSS Research Report no. 64, London: HMSO.

Roy, C. (1991) 'Les emplois du temps dans quelques pays occidentaux', *Données Sociales* 3: 223–5.

Rubery, J. (1996) 'The labour market outlook and the outlook for labour market analysis', in R. Crompton, D. Gallie and K. Purcell (eds) *Changing Forms of Employment: Organisation, Skills and Gender*, London: Routledge.

Rusega, S. and De Blas, A. (1985) *Mercado de trabajo y economia oculta en Andalucia*, Seville: Cuadernos IAR no. 3.

Sachs, I. (1984) *Development and Planning*, Cambridge: Cambridge University Press.

Sack, A.L. (1991) 'The underground economy of college football', *Sociology of Sport Journal* 8: 1–15.

Sada, P.O. and McNulty, M.L. (1981) 'The market traders in the city of Lagos', in P.O. Sada and J.S. Oguntoyinbo (eds) *Urbanisation Processes and Problems in Nigeria*, Ibadan: Ibadan University Press.

Safa, H.I. and Antrobus, P. (1992) 'Women and the economic crisis in the Caribbean', in L. Beneria and S. Feldman (eds) *Unequal Burden: Economic Crises, Persistent Poverty, and Women's Work*, Boulder, CO: Westview.

St Leger, F. and Gillespie, N. (1991) *Informal Welfare in Belfast: Caring Communities?*, Aldershot: Avebury.

Salmi, M. (1996) 'Finland is another world: the gendered time of homework', in E. Boris and E. Prugl (eds) *Homeworkers in Global Perspective: Invisible no More*, London: Routledge.

Santos, J.A. (1983) *A economia subterranea*, Coleccao estudos, serie A, no. 4, Lisbon: Ministrio do trabalho e seguranca social, .

Sanyal, B. (1991) 'Organising the self-employed: the politics of the urban informal sector', *International Labour Review* 130,1: 39–56.

Sassen, S. (1989) 'New York city's informal economy', in A. Portes, M. Castells and L. Benton (eds) *The Informal Economy: Studies in Advanced and Less Developed Countries*, Baltimore: Johns Hopkins University Press.

—— (1991) *The Global City: New York, London, Tokyo*, Princeton: Princeton University Press.

—— (1996) 'Service employment regimes and the new inequality', in E. Mingione (ed.) *Urban Poverty and the Underclass*, Oxford: Basil Blackwell.

Sassen, S. and Smith, R.C. (1992) 'Post-industrial growth and economic reorganisation: their impact on immigrant employment', in J. Bustamante, C.W. Reynolds and R.A. Hinojosa (eds) *US–Mexico Relations: Labour Markets Interdependence*, Stanford, CA: Stanford University Press.

Sassen-Koob, S. (1984) 'The new labour demand in global cities', in M.P. Smith (ed.) *Cities in Transformation*, Beverley Hills: Sage.

Sauvy, A. (1984) *Le Travail noir et l'economie de demain*, Paris: Calmann-Levy.

Schmitt, J. and Wadsworth, J. (1994) *Why are 2 Million Men Inactive? The Decline in Male Labour Force Participation in Britain*, Working Paper no. 338, London: Centre for Economic Performance.

Schneider, M. and Phelan, T. (1993) 'Black suburbanization in the 1980s', *Demography* 30: 269–79.

Schulz, M. (1995) 'The informal sector and structural adjustment: strengthening collective coping mechanisms in Tanzania', *Small Enterprise Development* 6,1: 4–14.

Scott, W.J. and Grasmick, H.G. (1981) 'Deterrence and income tax cheating: testing interaction hypotheses in utilitarian theories', *Journal of Applied Behavioural Sciences* 17,3: 395–408.

Sethuraman, S.V. (1981) *The Urban Informal Sector in Developing Countries*, Geneva: International Labour Office.

Sharpe, B. (1988) 'Informal work and development in the west', *Progress in Human Geography*, 12,3: 315–36.

Sik, E. (1994) 'From the multicoloured to the black and white economy: the Hungarian second economy and the transformation', *Urban Studies* 31,1: 47–70.

Simey, M. (1996) 'The end of work', *Local Work* 71: 1–6.

Simon, C.P. and Witte, A.D. (1982) *Beating the System: The Underground Economy*, Boston, MA: Auburn House.

Simon, P.B. (1997) *Crises of Urban Employment: An Investigation into the Structure and Relevance of Small-scale Informal Retailing in Kaduna, Nigeria*, mimeo.

Smith, J.D. (1985) 'Market motives in the informal economy', in W. Gaertner and A. Wenig (eds) *The Economics of the Shadow Economy*, Berlin: Springer-Verlag.

Smith, K. (1992) 'Reciprocity and fairness: positive incentives for tax compliance', in J. Slemrod (ed.) *Why People Pay Taxes*, Michigan: University of Michigan Press.

Smith, S. (1986) *Britain's Shadow Economy*, Oxford: Clarendon.

Smithies, E. (1984) *The Black Economy in England since 1914*, Dublin: Gill and Macmillan.

Standing, G. (1989) 'Global feminisation through flexible labor', *World Development* 17: 1077–96.

Stauffer, B. (1995) 'Regulation and reality: streetvending in Washington, DC', Paper presented at the *91st Annual Meeting of the Association of American Geographers*, Chicago, March.

Stepick, A. (1989) 'Miami's two informal sectors', in A. Portes, M. Castells and L. Benton (eds) *The Informal Economy: Studies in Advanced and Less Developed Countries*, Baltimore: Johns Hopkins University Press.

Stoleru, L. (1982) *La France à deux vitesses*, Paris: Flammarion.

Szinovacz, M. and Harpster, P. (1994) 'Couples' employment/retirement status and the division of household tasks', *Journal of Gerontology* 49,3: 125–36.

Tanzi, V. (1980) 'The underground economy in the United States: estimates and implications', *Banco Nazionale del Lavoro* 135: 427–53.

—— (1982) *The Underground Economy in the United States and Abroad*, Lexington, MA: D.C. Heath.

Tate, J. (1996) 'Making links: the growth of homeworker networks', in E. Boris and E. Prugl (eds) *Homeworkers in Global Perspective: Invisible no More*, London: Routledge.

Terhorst, P. and Van de Ven, J. (1985) *Zwarte Persoonlijke Dienstverlening en het Stedelijk Milieu*, Amsterdam: Social Geografisch Instituut.

Thomas, J.J. (1988) 'The politics of the black economy', *Work, Employment and Society* 2,2: 169–90.

—— (1992) *Informal Economic Activity*, Hemel Hempstead: Harvester Wheatsheaf.

Thomas, K. and Smith, K. (1995) 'Results of the 1993 Census of Employment', *Employment Gazette* 103: 369–84.

Thomas, R. and Thomas, H. (1994) 'The informal economy and local economic development policy', *Local Government Studies* 20,3: 486–501.

Thrift, N.J. (1996) *Spatial Formations*, London: Sage.

Tievant, S. (1982) 'Vivre autrement: échanges et sociabilité en ville nouvelle', *Cahiers de l'OCS*, vol. 6, CNRS, Paris.

Tokman, V.E. (1978) 'An exploration into the nature of informal–formal sector relationships: the case of Santiago', *World Development* 6 (September–October): 1065–75.

—— (1986) 'Adjustment and employment in Latin America: the current challenges', *International Labour Review* 125,5: 533–43.

Townsend, A.R. (1997) *Making a Living in Europe: Human Geographies of Economic Change*, London: Routledge.

Trundle, J.M. (1982) 'Recent changes in the use of cash', *Bank of England Quarterly Bulletin* 22: 519–29.

Turner, R., Bostyn, A.M. and Wight, D. (1985) 'The work ethic in a Scottish town with declining employment', in B. Roberts, R. Finnegan and D. Gallie (eds) *New*

Approaches to Economic Life: Economic Restructuring, Unemployment and the Social Division of Labour, Manchester: Manchester University Press.

US Congress Joint Economic Committee (1983) *Growth of the Underground Economy 1950–81*, Washington, DC: Government Printing Office.

US General Accounting Office (1989) *Sweatshops in New York City: A Local Example of a Nationwide Problem*, Washington, DC: US General Accounting Office.

Van Eck, R. and Kazemeier, B. (1985) *Swarte Inkomsten uit Arbeid: resultaten van in 1983 gehouden experimentele*, The Hague: CBS-Statistische Katernen nr 3, Central Bureau of Statistics.

—— (1990) *The Supply of Hidden Labour in the Netherlands: A Model*, The Hague: National Accounts Research Division, Central Bureau of Statistics.

Van Geuns, R., Mevissen, J. and Renooy, P.H. (1987) 'The spatial and sectoral diversity of the informal economy', *Tijdschrift voor Economische en Sociale Geografie* 78,5: 389–98.

Van Ours, J. (1991) 'Self-service activities and formal or informal market services', *Journal of Applied Economics* 23,3: 505–15.

Van Parijis, P. (1992) (ed.) *Arguing for Basic Incomes*, London: Verso.

—— (1996a) 'Basic income and the two dilemmas of the welfare state', *The Political Quarterly* 67,1: 57–8.

—— (1996b) 'L'allocation universelle contre le chomage: de la trappe au socle', *Revue Française des Affaires Sociales*, 50,1: 111–25.

Verschave, F-X. (1996) 'The House that Braudel built: rethinking the architecture of society', in OECD (ed.) *Reconciling Economy and Society: Towards a Plural Economy*, Paris: OECD.

Vinay, P. (1985) 'Family life-cycle and the informal economy in central Italy', *International Journal of Urban and Regional Research* 9: 82–98.

—— (1987) 'Women, family and work: symptoms of crisis in the informal economy of Central Italy', *Sames 3rd International Seminar Proceedings*, Thessaloniki: University of Thessaloniki.

Vogler, C. (1994) 'Money in the household', in M. Anderson, F. Bechhofer and J. Gershuny (eds) *The Social and Political Economy of the Household*, Oxford: Oxford University Press.

Walby, S. (1997) *Gender Transformations*, London: Routledge.

Waldinger, R. (1986) *Through the Eye of the Needle: Immigrants and Enterprise in the New York Garment Trade*, New York: New York University Press.

Waldinger, R. and Lapp, M. (1993) 'Back to the sweatshop or ahead to the informal sector', *International Journal of Urban and Regional Research* 17,1: 6–29.

Warde, A. (1990) 'Household work strategies and forms of labour: conceptual and empirical issues', *Work, Employment and Society* 4,4: 495–515.

Warde, A. and Hetherington, K. (1993) 'A changing domestic division of labour? Issues of measurement and interpretation', *Work, Employment and Society* 7,1: 23–45.

Warren, M.R. (1994) 'Exploitation or co-operation? The political basis of regional variation in the Italian informal economy', *Politics and Society* 22,1: 89–115.

Weatherley, R. (1993) 'Doing the right thing: how social security claimants view compliance', *Australian and New Zealand Journal of Sociology* 29,1: 21-39.

Weber, F. (1989) *Le Travail a côté: étude d'ethnographie ouvrière*, Paris: Institut National de la Recherche Agronomique.

Weck-Hanneman, H. and Frey, B.S. (1985) 'Measuring the shadow economy: the case of Switzerland' in W. Gaertner and A. Wenig (eds) *The Economics of the Shadow Economy*, Berlin: Springer-Verlag.

Weiss, L. (1987) 'Explaining the underground economy as social structure', *British Journal of Sociology* 38: 216–34.

Wenig, A. (1990) 'The shadow economy in the Federal Republic of Germany', in European Commission, *Underground Economy and Irregular Forms of Employment, Final Synthesis Report*, Brussels: Office for Official Publications of the European Communities.

Willatt, M. (1982) 'Italy's big black economy', *Management Today* September: 58–61.

Williams, C.C. (1996a) 'Local Exchange and Trading Systems (LETS): a new source of work and credit for the poor and unemployed?', *Environment and Planning A* 28,8: 1395–415.

—— (1996b) 'Local purchasing and rural development: an evaluation of Local Exchange and Trading Systems (LETS)', *Journal of Rural Studies* 12,3: 231–44.

—— (1996c) 'Local currencies and community development: an evaluation of green dollar exchanges in New Zealand', *Community Development Journal* 31,4: 319–29.

—— (1996d) 'An appraisal of Local Exchange and Trading Systems (LETS) in the United Kingdom', *Local Economy* 11,3: 275–82.

—— (1996e) 'The new barter economy: an appraisal of Local Exchange and Trading Systems (LETS)', *Journal of Public Policy* 16, 1: 55–71.

—— (1996f) 'Informal sector responses to unemployment: an evaluation of the potential of Local Exchange and Trading Systems (LETS)', *Work, Employment and Society* 10,2: 341–59.

—— (1997) *Reinvigorating the Local Economy in West Yorkshire*, London: Forum for the Future.

—— (1998) 'The "new localism" in local economic development: the rationales, strategies and potential for localising economies', Paper presented to the *Royal Geographical Society Annual Conference*, Kingston University, January.

Williams, C.C. and Thomas, R. (1996) 'Paid informal work in the Leeds hospitality industry: regulated or unregulated work?', in G. Haughton and C.C. Williams (eds) *Corporate City? Partnership, Participation and Partition in Urban Development in Leeds*, Aldershot: Avebury.

Williams, C.C. and Windebank, J. (1993) 'Social and spatial inequalities in the informal economy: some evidence from the European Community', *Area* 25,4: 358–64.

—— (1994) 'Spatial variations in the informal sector: a review of evidence from the European Union', *Regional Studies* 28,8: 819–25.

—— (1995a) 'Black market work in the European Community: peripheral work for peripheral localities?', *International Journal of Urban and Regional Research* 19,1: 23–39.

—— (1995b) 'Social polarisation of households in contemporary Britain: a "whole economy" perspective', *Regional Studies* 29,8: 723–8.

—— (1997) 'The unemployed and informal sector in Europe's cities and regions', in P. Lawless, R. Martin and S. Hardy (eds) *Unemployment and Social Exclusion: Landscapes of Labour Inequality*, London: Jessica Kingsley.

Windebank, J. (1991) *The Informal Economy in France*, Aldershot: Avebury.

—— (1996) 'To what extent can social policy challenge the dominant ideology of mothering? A cross-national comparison of France, Sweden and Britain', *Journal of European Social Policy* 6,2: 147–61.

Windebank, J. and Williams, C.C. (1995) 'The implications for the informal sector of European Union integration', *European Spatial Research and Policy* 2,1: 17–34.

—— (1997) 'What is to be done about the paid informal sector in the European Union? a review of policy options', *International Planning Studies* 2,3: 315–27.

Witte, A.D. (1987) 'The nature and extent of unrecorded activity: a survey concentrating on recent US research', in S. Alessandrini and B. Dallago (eds) *The Unofficial Economy: Consequences and Perspectives in Different Economic Systems*, Aldershot: Gower.

Woodward, R. (1995) 'Approaches towards the study of social polarization in the UK', *Progress in Human Geography* 19,1: 75–89.

Woody, B. (1992) *Black Women in the Workplace: Impacts of Structural Change in the Economy*, New York: Greenwood Press.

Wuddalamy, V. (1991) 'Les Mauriciens en France: une insertion socio-professionelle caracterisée par l'immigration spontanée', in S. Montagne-Villette (ed.) *Espaces et travail clandestins*, Paris: Masson.

Wyly, E.K. (1996) 'Race, gender and spatial segmentation in the twin cities', *The Professional Geographer* 48,4: 431–44.

Ybarra, J-A. (1989) 'Informalisation in the Valencian economy: a model for under-development', in A. Portes, M. Castells, and L.A. Benton (eds) *The Informal Economy: Studies in Advanced and Less Developing Countries*, Baltimore: Johns Hopkins University Press.

Zlolinski, C. (1994) 'The informal economy in an advanced industrialized society: Mexican immigrant labour in Silicon Valley', *Yale Law Journal* 103,8: 2305–35.

INDEX

Abrahamson, P. 138
active citizenship 161, 167, 187
Addison, T. 119
Aitken, S. 59
Alden, J. 12
Amado, J. 148
Amin, A. 29, 156
Amin, S. 133
Amott, T. 84
Anheier, H.K. 115, 123
Antrobus, P. 121
apparel industry 14, 88
Argentina 115
Atkinson, A. 154
Atkinson, A.B. 162
Australia 60, 100, 137, 165
Austria 100
Aznar, G. 159, 164

Bangladeshis 95
Barnes, H. 167
Barr, N. 149
barriers to entry 36–45, 58–61
barriers to eradication 140–4
Barthe, M.A. 23, 50, 62, 101, 108
Barthelemy, P. 29, 38, 40, 42, 63, 99
Baud, I.S.A. 120
Baxter, J. 72
Bayer, A.S. 16
Beatson, M. 150
Bebbington, K. 144
Beechey, V. 75
begging 120, 122
Belfast 24, 43, 60, 61–4, 68, 103, 108, 142
Belgium 52, 60, 79, 100, 137, 139
Bell, C. 75

Beneria, L. 39, 72, 121
Bennington, J. 138
Benton, L. 30, 32, 37, 39, 51, 103, 106, 107, 141
Berger, R. 50
Berthelier, P. 119
Berthoud, R. 84
Bhavnani, R. 84
Biggs, T. 151
Blair, J.P. 38, 50, 52, 98
Bloeme, L. 42
Bolivia 115
Bonneville, E. 59
Boris, E. 42, 87, 88, 122
Borocz, J. 152
Botswana 115
Brandt, B. 165
Brannen, J. 75
Brazil 115, 116
Briar, C. 72, 79
Briggs, V. 86
Brindle, D. 49
Bromley, R. 117
Brusco, S. 44
Bryson, A. 57
Bulgaria 41
Bunker, N. 101
Burawoy, M. 143
Button, K. 31, 98

Canada: Bas-du-Fleuve 45, 67, 71; Great North Peninsula 53; homeworking 79; measurements of informal employment 16, 24, 60, 66, 100; Montreal 45, 67, 71; Quebec 34–6, 45, 51, 54, 56, 57, 67, 71; social polarisation 137, 142; unpaid work 74, 76, 77

Cappechi, V.: on definitions 3, 32; on Emilia Romagna 107, 156; on gender and informal employment 71; on informal–formal relationship 30, 38, 39, 106, 175; on state and informal employment 141; on unemployed and informal employment 51
car repairs 45
Carbonetto, D. 33
cash–deposit ratios approach 16–19
cash-in-hand 15–16
Castells, M. 5, 29, 30, 49, 117, 133
catering industry 53
character of informal employment: by employment status 52-3, 119–20; by ethnicity 87–8; by gender 68–9, 120–2; by geography 101–3, 123–4; by developing nations 117–19
Chavdarova, T. 41
Cheng, L. 114, 118, 121, 185
child-care 27, 68, 73, 79
Chile 114, 115, 186
Chinese 84
Chu, Y.W. 41
citizen's income 161–3, 167–70
Citizen's Income Trust 162, 167
Clark, J. 139
Cobb, J.B. 161
Cocco, M.R. 17, 182
Cochrane, A. 139
Coffield, F. 59
Colombia 115, 118
Connolly, P. 117
Conroy, P. 153
construction industry 14, 68, 91
Contini, B. 17, 151
contract cleaning 108
Cook, D. 49, 59, 133, 183
corporatism 44
Cornuel, D. 16, 30, 40, 50, 53, 54, 58, 101, 107, 175
Costa Rica 115, 118
Costes, L. 90
Cousins, C. 79, 102
credit unions 164
crime 5, 15–16
Crompton, R. 183
Cubans 90–2, 93, 95
Culpitt, I. 138
cultural traditions 40–1, 63

Dagg, A. 79
Dallago, B. 21, 100
Daly, H.E. 161
Dasgupta, N. 117
Dauncey, G. 165
Davies, R.B. 186
Dawes, L. 154
De Blas, A. 53
De Grazia, R. 86, 152
De Klerk, L. 42, 63
De Pardo, M.L. 118
Deakin, S. 150, 154
Dean, H. 54, 59
Dean, P.N. 59
defining informal employment 1, 2–5
Del Boca, D. 12, 42, 43
Delphy, C. 183
Demery, L. 119
Denison, E. 13
Denmark: benefit recipient quotas 52; Copenhagen 45; magnitude of informal employment 100, 143; social polarisation 139; social security regulations 59; unpaid work 74, 77; Western Jutland 45; women's participation 66, 67
Denton, N.A. 84
deregulation of labour market 8, 134, 147–51
Desproges, J. 163
development paths 27–31
Dewberry, C. 101
Dex, S. 186
Dicken, P. 113
Dilnot, A. 21, 154
Dobson, R.V.G. 165
domestic service 14, 68, 69, 87, 120
Dore, R. 162
Dorling, D. 59, 186
Douthwaite, R. 159, 164
Duchrow, U. 161
Duncan, C.M. 30, 45
Duriez, B. 16, 30, 40, 50, 53, 54, 58, 101, 107, 175

East Asia 28, 84, 113, 114, 118
economic regulators of informal employment 38–40, 58, 61, 72–3, 93–4, 105–9, 124
Economist Intelligence Unit 50
ecologists 163

Ecuador 115
education 42
Ekins, P. 160, 163
elder care 27
electronics industries 14, 39
Elkin, T. 31, 98
Endres, C.R. 38, 50, 52, 98
Engbersen, G 52, 58
England: Barnsley 18; Birkenhead 18; Brighton 18; Brixton 95; Calderdale 165; Cleveland 54; Hartlepool 70, 108; Isle of Sheppey 23–4; Kirkcaldy 55; London 18, 54, 184, 185; Luton 54; Manchester 165; north-east 54; Oldham 95; south-west 55, 108; Totnes 165; Walsall 18; West Yorkshire 171
environmental regulators of informal employment 44–5, 64, 96, 106–9, 126
Espenshade, T.J. 86
Esping-Andersen, G. 149, 150, 153
ethnic enclaves 106
Etzioni, A. 160
European Commission 28, 85, 99, 100, 132, 136, 139, 153, 160, 185
Eurostat 73, 136
Evason, E. 16, 24, 54, 57, 154
Evers, A. 163

Fainstein, N. 84, 96, 153
Family Expenditure Survey 20–1
Farley, R. 84
fashion industry 103, 120
Fashoyin, T. 116, 117
Feige, E.L. 5, 19, 39, 132
Feldman, S. 45
Felt, L.F. 3, 53
Ferman, L. 33, 45, 60
Fernandez-Kelly, M.P. 14, 83, 84, 92, 93
Filani, M.O. 126
financial exclusion 16, 18
Finland 76, 79
flea markets 33
Floro, M. 122
formalisation thesis 27–9, 46, 113–16, 174–6, 178
Forte, F. 12, 42, 43
Fortin, B.: on attitudes towards informal employment 142; on character of informal employment 32, 33, 34; on education levels 42; on gender 66, 68, 70, 71; on geography 45; on informalisation thesis 30, 175; on methods 22, 23, 24; on unemployed 51, 52, 54, 56, 57, 58
Forum for the Future 170–1, 187
Foudi, R. 50, 55, 62, 101, 108
France: attitudes towards tax and benefit fraud 60; Auvergne 40; commuter villages 54; definitions 2; employment 44; geography 101, 107, 108; Grand Failly 39, 64; immigrant informal workers 90; immigration 86; Limousin 40; magnitude of informal employment 100; Poitou-Charente 40; social polarisation 137; unemployed and informal employment 49, 50; unpaid work 74, 76, 77, 79
Frank, A.G. 29, 117, 133
Frechette, P. 34
Freud, D. 15
Frey, B.S. 19, 22, 39, 41
Friedmann, Y. 159
fruit picking 108
full-employment, critique of 27–9, 135–7
furnituremaking industries 120

Gabor, I.R. 152
Gallie, D. 62
Galster, G.C. 84, 184
garbage picking 122
Garcia, A.M. 14, 83, 84, 92, 93
garment industries 14, 23
Gass, R. 159
Geeroms, H. 39
gender divisions of domestic labour 73–8, 183–4
geography of informal employment: cross-national comparisons 99–100; local and regional variations 101–3; typology of localities 104–9
Gereffi, G. 114, 118, 185
Germany: attitudes toward tax and benefit fraud 60; definitions 2; gender 68, 71–2; immigration 86, 94; magnitude of informal employment 100; non-wage labour costs 40; social polarisation 137; unemployed and informal employment 50, 56, 59; unpaid work 78

Gerry, C. 117
Gershuny, J. 3, 27, 66, 75, 76, 77, 183
Ghana 31, 112
Ghavamshahidi, Z. 121
Giddens, A. 33
Gilbert, A. 120, 156
Gilder, G. 149, 150
Gillespie, N. 41
Ginsburgh, V. 42
Glatzer, W. 50
Glennie, P.D. 167
Gornick, J.C. 72
Gorz, A. 159, 161
Grasmick, H.G. 39
grass-roots initiatives 163–7, 170–1
Gray, J. 149
Greco, T.H. 165
Greece 51, 52, 60, 86, 103, 106, 137
Greffe, X. 164
Gregg, P. 137
Gregory, A. 66, 73
Gregson, N. 79
Gringeri, C. 44, 79
Grosskoff, R. 185
Grossman, G. 152
Guisinger, S. 118
Gutmann, P.M. 17, 30, 39, 49, 50, 112, 132

Hadjicostandi, J. 121
Hadjimichalis, C. 45, 51, 60, 86, 98, 103, 106
Hahn, J. 122
hairdressing 45
Haitians 90–2, 93, 95
Hakim, C. 66
Hall, P. 114
Hanson, S. 72
Harding, P. 3, 24, 27, 182
Harpster, P. 75
Harris, J.R. 122
Hart, K. 31, 112, 117, 126, 185
Hasseldine, D. 144
Haughton, G. 98
Heinze, R.G. 161, 163, 167
Hellberger, C. 50, 52, 56, 59, 60, 68, 71, 72
Henry, J. 15
Henry, S. 18, 33, 55, 112
Herbert, A. 95
Hessing, D. 54
Hetherington, K. 66

Hewitt, P. 186
high denomination notes method 15–16
HM Treasury 24
Holzer, H. 84
home repairs 45, 53, 68, 69
homeworkers 23, 42–3, 44, 53, 79, 87, 92, 93, 103, 120–1
Honduras 114, 115
Hong Kong 113, 114
Houghton, D.18
housing tenure 45
housekeepers 56
Houston, J.F. 39
Howe, L.: on geography 108; on methods 23; on skills 38; on social regulators 37, 40, 142; on unemployed 50, 52, 54, 57, 58, 60, 63
Hoyman, M. 183
Hsuing, P.C. 121
Huber, E. 123, 186
Hungary 76, 77, 152
Hutton, W. 150

immigration policy 94
income/expenditure discrepancies approach 20–2
Indians 90
Indonesia 114
industrial structure 38
informal credit schemes 95
informalisation thesis 1, 7, 29–30, 46, 113–16, 174–6, 178
institutional regulators of informal employment 42–4, 64, 78–80, 94, 105–9, 125–6
International Labour Organisation 29, 72, 113, 114, 115, 117, 123, 126, 132, 185
Iran 121
Ireland 78, 100, 137
Irfan, M. 118
Isachsen, A.J. 19, 22, 23, 49
Isuani, E.A. 123
Italy: definitions 2; deregulation 156; Emilia Romagna 39; gender and informal employment 66, 67, 71, 79, 86; geography of informal employment 101, 102–3, 106, 107; magnitude of informal employment 100; non-wage labour costs 40; social

polarisation 137, 139; unemployed and informal employment 51, 52, 56, 57; union power 44.
Izquierdo, M.J. 53

Jackson, K.T. 84
Jacobs, J. 72
Jacobs, J.A. 57
Japanese 84
Jarvis, H. 186
Jencks, C. 184
Jenkins, R. 3, 24, 27, 182
Jensen, L. 32, 51, 53, 54, 57, 68, 85
Jessen, J. 41, 45, 54, 57, 103
Jobseekers' Allowance 49
Johnson-Anumonwo, I. 84
Jones, P.N. 86, 93
Jones, T. 84
Jordan, B. 54, 55, 57, 59, 108, 162, 184
Jung, Y.H. 39

Kalinda, B. 122
Kasarda, J.D. 84, 184
Kazemeier, B. 50, 56, 57, 58, 66, 67
Keenan, A. 59
Kempson, E. 95
Kesteloot, C. 60
Kiernan, K. 72, 75, 77, 78
Kinsey, K. 144
kinship networks 41
Klovland, J.T. 39
Komter, A.E. 40, 53, 54, 58, 107, 142
Koopmans, C.C. 50
Korbonski, A. 28
Kroft, H.G. 40
Kumar, K. 27

labour legislation avoidance 4, 14, 42, 43
Lagos, R.A. 31, 116, 117
Laguerre, M.S. 32
Lang, P. 165
Lapp, M. 32
Latin America 29, 86
Lautier, B. 116, 117, 120, 122, 123, 125
Laville, J.-L. 159, 164
lax enforcement of legislation 141
Lee, R. 160
Legrain, C. 40, 41, 63, 64, 107

Lemieux, T. 22, 34, 36, 42, 51, 52, 66, 71
Leonard, M.: on character of informal employment 32, 53, 54, 57; on ethnicity 89; on gender 68, 70; on informal–formal relationship 30, 175; on institutional regulators 43; on methods 22, 23, 24; on social regulators; 40, 41, 142; on unemployed 60, 61–4
Leontidou, L. 60, 103, 106
LETS 164–7, 170, 187
Levitan, L. 45
Lewis, A.185
Leyshon, A. 16, 166
Lim, I.Y.C. 88, 121
Lin, J. 23, 32, 83, 84, 87, 88, 90
Lindbeck, A. 149, 150
Linton, M. 165
Lipietz, A. 158, 161, 168, 169
Lobo, F.M. 13, 43, 44, 51, 52, 53, 60, 67, 103, 105, 106, 141
local food links 164
Lomnitz, L.A. 134, 152
Lopez, C. 53
Lowe, M. 79
Lozano, B. 33, 51
Lubell, H. 116, 117, 120
Lukacs, J. 143
Luxembourg 78, 137, 139
Lysestol, P.M. 41, 54, 55, 58, 59

Mabogunje, A.L. 126
Macafee, K. 21
McCrohan, K. 22
MacDonald, R. 24, 52, 54, 55, 57, 66, 70, 132, 133
Macfarlane, R. 163
McGee, R.T. 39
McInnis-Dittrich, K. 66, 68, 71
McKee, L. 75
McLafferty, S. 84
McLaren, D. 31, 98
McLaughlin, E. 154
McNulty, M.L. 126
Magatti, M. 51, 87, 132, 134
magnitude of informal employment: by employment status 34, 50–2; by ethnicity 83–4; by gender 34, 66–7; by geography 45, 99–103; by household income 34; by immigration 85–7

Maguire, K. 143
Main, B.G.M. 55
Malaysia 114
Maldonado, C. 31, 116, 117
Malone, A. 49
marginality thesis 1, 7, 31–2, 46,
    116–24, 176–7, 178
Marsden, T. 167
Martin, C.J. 66
Martin, J. 116, 120, 123
Martino, A. 86, 152
Massey, D. 73, 104, 110
Massey, D.S. 84
Mathur, O.P. 123
Mattera, P. 13, 18, 21, 56, 71, 85, 86,
    101
Matthaei, J. 84
Matthews, K.G.P. 17, 18, 39, 49, 147,
    151, 155
Mauritius 115
Mayer, S. 184
Mayo, E. 158, 159, 160, 161, 162, 163
Meade, J.E. 162
Meadows, T.C. 17
Meagher, K. 117
measuring informal employment: 7;
    indirect methods 12–22, 99; direct
    methods 22–4
meat processing industry 14
Meehan, E. 138
Meert, H. 41, 42, 45, 60
Melgar, A. 185
Melrose, M. 54, 59
metal industries 120
Mexicans 93
Mexico 86, 115, 116, 121
Micklewright, J. 154
Middle East 29, 115
Miguelez, F. 60
Miles, I. 58, 165
Minc, A. 147, 148
Mingione, E.: on character of informal
    employment 32; on civil society 154;
    on economic regulators 38; on
    gender 66, 67; on geography 101,
    102, 103, 104, 106, 107; on
    immigration 87; on institutional
    regulators 43; on regulating informal
    employment 132, 134, 135, 136,
    141; on social regulators 41, 43; on
    unemployed 51, 57, 58
Miraftab, F. 120, 123

Mirus, R. 16
Modood, T. 84
Mogensen, G.V. 30, 45, 59, 60, 66, 67,
    143, 175
Momsen, J.H. 120
monetary transactions approach 19–20
Mont, J. 39
Morlicchio, E. 51, 101
Morris, C.N. 21
Morris, L.: on gender 70, 186; on
    geography 108; on housing tenure
    45; on informal–formal relations 30,
    175; on social networks 37, 41, 165;
    on unemployed 50, 52, 54, 55, 57,
    58, 59, 62
Moser, C. 123
Moss, P. 75
motivations for working informally: by
    employment status 53–5; by ethnicity
    88–9; by gender 69–70
Moulier-Boutang, Y. 86, 87
Murray, C. 149
Myles, J. 140, 151, 154

Nattrass, N.J. 126
Netherlands: citizen's income 169;
    definitions 2; gender 66, 67;
    geography 101, 107, 109; homework
    44; magnitude of informal
    employment 100; social polarisation
    137, 142; unemployed 50, 52, 54,
    56, 57, 58, 62, 64; unpaid work 74,
    76, 77
new economics 8, 152–72
New Economics Foundation 158, 186
new international division of labour 113
New Zealand 165
Nicaise, I. 28, 136
Nicaraguans 90–2, 95
Nigeria 115, 119
Noble, M. 18, 21, 38, 39
North Africa 29, 115
Norway 41, 59, 74, 76, 77, 100
Nurulk Amin, A.T.M. 126
Nyssens, M. 164

Oakley, A. 139
OECD 132, 137, 153, 158, 160, 183
Offe, C. 161
O'Higgins, M. 18, 21, 38, 39
Okun, A.M. 149, 150
Oliver, M. 84

Olk, T. 163, 167
OPCS 27
Owen, D. 84

Pacione, M. 165
Paglin, M. 20
Pahl, R.E.: on barriers to participation 155, 165; on character of informal employment 32; on economic regulators 38; on full-employment 28, 135–6; on gender 66, 69; on housing tenure 45; on informal–formal relationship 30, 175; on methods 22, 23; on social polarisation 186; on unemployed 49, 50, 52, 54, 57, 58, 59, 64; on unpaid work 75
Pakistani community 95
Panama 114, 115
Paraguay 114, 115
Paredes-Cruzatt, P. 118, 120, 121
Parker, H. 49, 162
participative planning 164
pay rates: 34–6; in developing nations 118–19; by employment status 34, 55-7; by ethnicity 89–90; by gender 34, 70–2; by geography 102–3
Peacock, A.T. 186
Peattie, L.R. 117
Peck, J. 104, 133, 143, 148, 150, 151, 153, 154
Perkins, T. 75
personal services 14, 68
Peru 115, 118
Pestieau, P. 42, 60, 63, 100
Peterson, H.G. 49
Phelan, T. 184
Phizacklea, A. 23, 42, 68, 87, 88
Pihera, J.A. 17
Pinch, S. 75, 150, 153, 186
Polanyi, K. 150, 185
Porritt, J. 163
Porter, R.D. 16
Portes, A.: on definitions 4, 5; on character of informal employment 33, 117, 118; on ethnicity and immigration 83, 84, 85, 87, 88, 90, 91; on informalisation 29, 30; on institutional regulators 43, 44, 106; on measuring informal employment 14–15, 22; on regulating informal

employment 133, 134, 141, 143; on social regulators 42; on unemployed 49, 64
Portugal 13, 44, 52, 53, 67, 100, 137, 139, 182
Pratt, G. 72
Preston, V. 84
Priest, G.L. 66
printing industry 120
Prugl, E. 42, 68, 87, 122
Psacharopoulos, G. 122
Pugliese, E. 87, 88, 90

Rakowski, C.A. 117, 124
Rastogi, A. 17, 18
Recio, A. 60, 141
Redley, M. 55, 162
Rehfeldt, U. 138
Reissert, B. 52, 139, 145
Renooy, P.: on character of informal employment 32, 165; on gender 66, 69; on geography 101, 107, 109; on housing tenure 45; on institutional regulators 39; on social regulators 41, 42; on unemployed 54, 56, 57, 58, 63, 64
Republic of Korea 113
restaurants 14
retail industry 53
retired persons 56
Richardson, H.W. 117, 122, 126
Rifkin, J. 159
Roberts, B. 33, 116, 118, 120, 123, 124, 126, 133
Roberts, C. 66, 84
Robertson, J. 159, 161, 163, 164
Robson, B.T. 50, 98
Roldan, M.I. 39, 182
Rosanvallon, P. 31, 49
Rosewell, B. 161
Rostow, W.J. 27, 113
Roth, J. 133
Roustang, G. 164
Rowlingson, K. 54, 55, 132, 133
Roy, C. 74
Rubery, J. 150, 153, 183
rural areas 44–5, 51
Rusega, S. 53

Sachs, I. 160, 164
Sack, A.L. 89

Sada, P.O. 126
Safa, H.I. 121
Salmi, M. 79
Santos, E. 17
Santos, J.A. 182
Sanyal, B. 117
Sassen, S.: on character of informal employment 32; on definitions 5; on economic regulators 38; on geography 107; on global cities 113; on immigration and ethnicity 83, 84, 87, 88, 90; on methods 14; on unemployed 49, 63
Sassen-Koob, S.: on economic regulators 37; on education 42; on geography 106; on immigration and ethnicity 85, 91; on institutional regulators 43, 44; on methods 14; on unemployed 64
Sauvy, A. 147, 148, 151, 152
Schmitt, J. 45
Schneider, M. 184
Schulz, M. 122
Schwarze, J. 50, 52, 56, 59, 60, 68, 71, 72
Scotland 41
Scott, W.J. 39
segmented informal labour market 32–6, 64
self-employment 12–3, 20, 38
Sethuraman, S.V. 122
settlement size 44–5
Shapiro, T.M. 84
Sharpe, B. 117
Shaw, G.K. 186
shoe industry 120
Sik, E. 28, 143
Simey, M. 163
Simon, C.P. 45, 49
Simon, P.B. 119, 120, 123, 124, 126
Sinclair, P.R. 3, 53
Singapore 113, 114
skills 58, 60
Smith, J.D. 17, 19, 22
Smith, K. 28, 136
Smith, R.C. 14
Smith, R.S. 16, 144
Smith, S. 18, 20, 21, 58, 59, 165, 166, 182
Smithies, E. 18, 104
Social Charter 138–40
social democrats 163, 164

social isolation 62, 63
social networks 41, 58, 60, 62–3, 70, 155
social norms and moralities 40–1, 62, 63
social polarisation of households 137–8, 153–4
social regulators of informal employment 40–2, 58, 62–3, 73–8, 95–6, 106–9, 125–6
social security fraud 1, 4, 43, 49, 54, 60, 63, 132–3, 143, 183
socio-economic mix 41, 63
Soto, H. de 29, 125, 126, 147, 148, 151, 153
South East Asia 28, 113, 139
South Korea 114
Spain: flexible labour markets 79; gender 67; geography 103, 106, industrial homeworking 42; magnitude 100, non-wage labour costs 40; 107, 110, social polarisation 137; unemployed 51, 60
spatial divisions of informal labour 101–4
St Leger, F. 41
Standard Industrial Classification (SIC) index 3
Standing, G. 120
state regulations 43–4, 64
Stauffer, B. 50, 87
Stepick, A. 83, 84, 87, 88, 90, 91
Stimpson, C.R. 72
Stoffaes, C. 148
Stoleru, L. 148
Storey, A. 75
Strom, S. 49
students 56
sub-contracting 39
sub-Saharan Africa 29
Sweden 60, 100
Szinovacz, M. 75

Taiwan 114, 121
Tanzi, V. 16, 17, 18, 39
Tate, J. 79
tax evasion 1, 4, 38, 39–40, 60, 63, 132–3, 144, 183
tax levels 39
Taylor, M. 186
Terhorst, P. 42, 63
Thailand 114

Third World employment 28–9
Thomas, H. 39
Thomas, J.J.: on definitions 2, 3, 5; on
  gender 120, 121, 182; on
  immigration 86, 185; on magnitude
  29; on measurements 18, 20, 21, 22,
  23, 182; on pay rates 118; on
  unemployed 58, 165
Thomas, K. 28, 136
Thomas, R. 39
Thrift, N. J. 16, 33, 166, 167
Tievant, S. 50, 101
Todaro, M. 122
Tokman, V.E. 117, 118
Townsend, A.R. 68, 72, 136
Trundle, J.M. 16, 17
Tunisians 90
Turner, R. 41, 63
Tzannatos, Z. 122

United Kingdom: attitudes toward tax
  and benefit fraud 54, 60;
  deregulation 148, 150; ethnicity
  and immigration 84; gender 66;
  geography 101–2; homeworking 79;
  LETS 165; magnitude of informal
  employment 100; measuring 16, 18,
  20; 'shop a dole cheat' hotline 49;
  social polarisation 137, 153, 154;
  social policy 139, unemployed 49,
  50, 57, 60; unpaid work 74, 76, 77
unlicensed taxi-driving 108
unpaid informal work 4, 27, 73–8
urban areas 45
Uruguay 118, 185
USA: attitudes toward tax and benefit
  fraud 60; Congress Joint Economic
  Committee 13; deregulation 148,
  150–1; ethnicity and immigration 82,
  83, 84, 85, 86, 89, 90–2, 96; gender
  66, 68; General Accounting Office
  14; geography 103, 106;
  homeworking 79; Los Angeles 92,
  93; measurements 13, 14, 15, 16, 17,
  20, 100; Miami 90–2, 95; New York
  88, 91; North California 33;
  Pennsylvania 51; San Francisco 51;
  social polarisation 137, 153, 154;
  unemployed 50, 52, 57, 60;
  unpaid work 74, 76, 77; Washington
  91

Vaiou, D. 45, 51, 60, 86, 98, 103, 106
Van de Ven, J. 42, 63
Van Eck, R. 50, 56, 57, 58, 66, 67
Van Geuns, R.: on economic regulators
  38; on environmental regulators 45;
  on geography 101, 103, 105, 106,
  107; on institutional regulators 43,
  141; on social regulators 40, 41, 42;
  on unemployment 50
Van Ours, J. 45
Van Parijis, P. 162
Venezuela 115
Verschave, F.-X. 160
very small enterprises 14, 38
Vijgen, J. 42, 63
Vinay, P. 44, 66, 71, 79, 141
Vogler, C. 66

Wadsworth, J. 45, 137
Walby, S. 66, 72, 73, 75
Waldinger, R. 32, 88
Wales 70
Warde, A. 22, 24, 41, 50, 66
Warren, M.R. 5, 32, 37, 43, 44, 51, 64,
  106, 141, 156
Weatherley, R. 59, 60
Weber, F. 40, 41, 63, 103
Weck, H. 19, 22, 39, 41
Weck-Hanneman, H. 22
Weiss, L. 37
welfare benefit regulations 43
welfare state, critiques 138–40, 148–52
Wenig, A. 43, 59, 60, 63, 64, 132
Wilkinson, F. 150, 154
Willatt, M. 86
Williams, C.C.: on assisted self-help
  164; on definitions 2; on geography
  101, 105; on LETS 165, 166, 187;
  on localisation 171; on policy
  towards informal employment 182;
  on social polarisation 153, 186; on
  sub-contracting 39; on unemployed
  53, 155, 166; on welfare state 139
Williams, S. 139
Windebank, J.: on definitions 2; on
  gender 66, 73; on geography 101,
  105; on housing tenure 45; on policy
  towards informal employment 182;
  on social polarisation 153, 186; on
  unemployed 53, 155, 166; on unpaid
  work 27, 79; on welfare state 139

Wintersberger, H. 163
Witte, A.D. 38, 45, 49
Wolkowitz, C. 23, 42, 68, 88
Woods, R. 16, 24, 54, 57, 154
Woodward, R. 59, 186
Woody, B. 84
Wrigley, N. 167

Wuddalamy, V. 86, 88
Wyly, E.K. 84

Yaounde 119
Ybarra, J.-A. 29, 133

Zlolinski, C. 87